AF587516

Measurement of Multicomponent Diffusion in Liquids Using Raman Microspectroscopy and Microfluidics

Messung von Multikomponentendiffusion in Flüssigkeiten mit Raman-Mikrospektroskopie und Mikrofluidik

Von der Fakultät für Maschinenwesen der
Rheinisch-Westfälischen Technischen Hochschule Aachen
zur Erlangung des akademischen Grades
einer Doktorin der Ingenieurwissenschaften
genehmigte Dissertation

vorgelegt von

Christine Jutta Peters geb. Blesinger

Berichter: Univ.-Prof. Dr.-Ing. André Bardow
Directeur de recherche au CNRS, Jean-Baptiste Salmon, Ph. D.

Tag der mündlichen Prüfung: 27. Mai 2019

Diese Dissertation ist auf den Internetseiten der Universitätsbibliothek online verfügbar.

Aachener Beiträge zur Technischen Thermodynamik Band 25

Christine Jutta Peters
Measurement of Multicomponent Diffusion in Liquids Using Raman Microspectroscopy and Microfluidics

Messung von Multikomponentendiffusion in Flüssigkeiten mit Raman-Mikrospektroskopie und Mikrofluidik

ISBN: 978-3-95886-337-8

Bibliografische Information der Deutschen Bibliothek
Die Deutsche Bibliothek verzeichnet diese Publikation in der Deutschen Nationalbibliografie; detaillierte bibliografische Daten sind im Internet über http://dnb.ddb.de abrufbar.

Herstellung & Vertrieb:

1. Auflage 2020

Süsterfeldstr. 83, 52072 Aachen
Tel. 0241 / 87 34 34 00
www.Verlag-Mainz.de

ISSN: 2198-4832

Satz: nach Druckvorlage des Autors
Umschlaggestaltung: Druckerei Mainz

printed in Germany
D82 (Diss. RWTH Aachen University, 2019)

Vorwort

Die vorliegende Arbeit entstand während meiner Tätigkeit als wissenschaftliche Mitarbeiterin am Lehrstuhl für Technische Thermodynamik an der RWTH Aachen. Zuallererst möchte ich mich bei Prof. André Bardow für das in mich gesetzte Vertrauen und die daraus resultierenden Möglichkeiten, die erfolgreiche Zusammenarbeit und die konstruktiven Diskussionen danken. Jean-Baptiste Salmon danke ich für die übernahme des Koreferats und den spannenden fachlichen Austausch. Prof. Schmitt danke ich für die übernahme des Prüfungsvorsitzes und die angenehme Art, die Prüfung zu leiten.

Ganz herzlich bedanken möchte ich mich auch bei allen aktiven & ehemaligen Mitarbeiterinnen und Mitarbeitern des LTT: Es war eine tolle Zeit, in der ich viel gelernt habe! Mein besonderer Dank gilt meinen "Büromitbewohnerinnen" Dominique & Julia, für die fachlichen Diskussionen, den Rat, die geteilte Schokolade und die gemeinsame Zeit. Ebenso möchte ich mich bei Peter, Bastian, Carsten, Ludger & allen anderen für den fachlichen Austausch und den gemeinsamen "Freizeitstress" bedanken. Meinen Gruppenleitern Paule, Toddy & Hans-Jürgen danke ich für ihre Unterstützung und ihr offenes Ohr. Unbedingt bedanken möchte ich mich auch bei Willi und allen TaMis für ihre unermüdliche Unterstützung. Mein Dank gilt auch allen Studierenden, die mich als stud. Hilfskraft oder im Rahmen ihrer Abaschlussarbeiten begleitet haben, insbesondere Rafael, Julia, Sandra, Gwen, Alex & Andy.

Stellvertretend für alle Kolleginnen & Kollegen in der akademischen Selbstverwaltung danke ich Wolfgang, Mathias und Claudia für die erfolgreiche Zusammenarbeit und das in mich gesetzte Vertrauen.

Nicht zuletzt danke ich meinen Eltern, meinen Schwiegereltern, meinen Freunden & Lars - für ihre Unterstützung, ihren Rat und tolle gemeinsame Zeiten.

Christine

Contents

List of Figures

List of Tables

Notation

Roman Symbols

A	channel half width	m
A	integrated spectral area	a.u.
A	parameter in the excess volume Redlich-Kister type function	$m^3\,mol^{-1}$
A	factor in Arrhenius type temperature correlation	$m^2\,s^{-1}$
$\boldsymbol{B}$	transformation matrix	
$\boldsymbol{B}$	inverse of Maxwell-Stefan diffusion coefficient matrix	$s\,m^{-2}$
B	parameter in the excess volume Redlich-Kister type function	$m^3\,mol^{-1}$
B	exponent in Arrhenius type temperature correlation	K
b	weight of component models in IHM procedure	
C	spiral radius	m
C	parameter in the excess volume Redlich-Kister type function	$m^3\,mol^{-1}$
c	molar concentration	$mol\,m^{-3}$
$\boldsymbol{D}$	diffusion coefficient matrix	$m^2\,s^{-1}$
$Đ$	Maxwell-Stefan diffusion coefficient	$m^2\,s^{-1}$
d	diameter	m
De	Dean number	
e	exponent	
f	body forces	$kg\,m^{-2}\,s^{-2}$
Fo	Fourier number	
G	Gibbs energy	J

g	gravitational acceleration	$\mathrm{m\,s^{-2}}$
h	height	m
H	channel height	m
J	diffusive flux	$\mathrm{mol\,m^{-2}\,s^{-1}}$
k	Boltzmann constant	$\mathrm{J\,K^{-1}}$
k	coverage factor	
k	interaction parameter in PC-SAFT	
k	spectral calibration factor	
L	channel length	m
l	length	m
M	molar mass	$\mathrm{kg\,mol^{-1}}$
m	exponent/factor of parabolic velocity profile	
n	number	
n	exponent in Tyn eq., depending on heat of vaporization	
o	index of observation points	
p	index of Raman spectra at observation point s	
Pe	Peclet number	
R	gas constant	$\mathrm{J\,mol^{-1}\,K^{-1}}$
R	reaction rate	$\mathrm{mol\,m^{-3}\,s^{-1}}$
r	molecular radius	m
r	correlation coefficient	
Re	Reynolds number	
p	pressure	Pa
s	axial coordinate and observation point	m
Sc	Schmidt number	
T	absolute temperature	K
t	temperature	°C
t	time	s
u	relative error	%
V	molar volume	$\mathrm{m^3\,mol^{-1}}$

$\bar{V}$	partial molar volume	$m^3\,mol^{-1}$
$\dot{V}$	volume flow	$m^3\,s^{-1}$
v	velocity	$m\,s^{-1}$
$\bar{v}$	average velocity	$m\,s^{-1}$
W	channel width	m
w	width	m
x	mole fraction	mol %
y	width coordinate	m
Y	dimensionless position of the interface	
z	height coordinate	m

Greek Symbols

α	factor to determine critical Reynolds number	
β	inlet angle	°
$\boldsymbol{\Gamma}$	thermodynamic factor	
γ	activity coefficient	
δ	Kronecker delta	
δ	width of diffusion zone	m
ϵ	association strength	$J\,mol^{-1}$
η	viscosity	mPa s
κ	effective volume of an association site	
κ	thermal diffusivity	$m^2\,s^{-1}$
λ	wavelength	m
μ	chemical potential	$J\,mol^{-1}$
μ	experiment	
μ	mean	
ν	relative wavenumber	m^{-1}
Φ	IHM component model	
ϕ	association factor	

ρ	density	$kg\,m^{-3}$
$\sum$	unity of mole fractions	
$\sigma(D)$	standard deviation of diffusion coefficient	$m^2\,s^{-1}$
$\sigma(x)$	standard deviation of mole fractions	mol %
$\sigma(x_{\mathrm{meas}} - x_{\mathrm{calc}})$	standard deviation of fitting error	mol %
τ	time	s
τ	contact time	s
ζ	inversion width of interface	m
Ψ	IHM borosilicate glass model	
Ω	IHM mixture model	

Superscripts

α	phase α
β	phase β
E	excess
M	molar reference frame
V	volume reference frame
0	pure component
†	eigenvalue
⋆	geometric mean

Subscripts

A	component A
B	component B
c	n-th component
c	critical
curvature	curvature of a spiral or meander

cal	spectral calibration
calc	calculated
e	thermal entrance length
e	time to steady state
evaluations	evaluations of fitting procedure
f	entrance flow region
fit	fitting procedure
H	hydraulic
i	component
i	inner
in	flowing in
j	component
j	observation point
k	component
l	left
max	maximal
meas	measured
min	minimal
mix	mixture
opt	optimal
pred	prediction
R	Raman (spectra)
r	right
r	relative
opt	optimal
out	flowing out
s	observation points
s	single evaluation
self	self-diffusion coefficient
spectra	spectra

t	total
weigh	prepared by weight
0	initial

Abbreviations and Acronyms

a.u.	arbitrary units
CCD	charge coupled device
corr	correlation
COSMO-RS	conductor like Screening Model for Real Solvents
Darken-KvB	multicomponent Darken model by Krishna, and van Baten
Darken-LBV	multicomponent Darken model by Liu, Bardow, and Vlugt
DKB	Darken-type multicomponent extension to Vignes eq. by Krishna, and van Baten
DLS	dynamic light scattering
eqn	equation
HPLC	high pressure liquid chromatography
IHM	indirect hard modeling
KT	multicomponent extension to Vignes eq. by Kooijman, and Taylor
LLE	liquid-liquid equilibrium
max.	maximum
MD	molecular dynamics
MS	Maxwell-Stefan
M.Sc.	Master of Science
μTAS	micro total analysis systems
NA	numerical aperture
Nd:YAG	neodymium-doped yttrium aluminum garnet (Nd:$Y_3Al_5O_{12}$)
NRTL	non-random two-liquid model
EOS	equation of state
PEEK	polyether ether ketone

PLS	partial least squares
PTFE	polytetrafluoroethylene
PC-SAFT	Perturbed-Chain Statistical Associating Fluid Theory equation of state
RMSE	root mean square error
RS	multicomponent extension to Vignes eq. by Rehfeldt, and Stichlmair
tr	trace
UV/Vis	ultraviolet–visible spectroscopy
VKB	Vignes-type multicomponent extension to Vignes eq. by Krishna, and van Baten
VLE	vapor-liquid equilibrium
Vignes-LBV	multicomponent extension to Vignes eq. by Liu, Bardow, and Vlugt
WCA	Weeks-Chandler-Andersen potential
WD	working distance
WK	multicomponent extension to Vignes eq. by Wesselingh, and Krishna

List of Student Research Projects

This dissertation contains material that has been elaborated by students in form of research projects, bachelor's theses, and master's theses. The following list gives details.

- Frederic Buttler, "Analysis of the Temperature Dependency of Fick Diffusion Coefficients in Liquids", diploma thesis, 2013.
- Alexandra Juring, "Validierung eines optimierten mikrofluidischen Messverfahrens für die Bestimmung von multikomponenten Diffusionskoeffizienten in Flüssigkeiten", research project, 2013.
- Andrea Rathgeb, Astrid Sommer, Bernhard Nikolic, "Untersuchung des Temperatureinflusses auf die Diffusion in binären Flüssigkeiten", research project, 2013.
- Sascha Deric, Nils Hammesfahr, Dominik Tillmanns, "Experimentelle Untersuchung von temperaturabhängigen Viskositäten und Dichten in binären Gemischen aus ionischen Flüssigkeiten und Wasser", research project, 2013.
- Adrian Caspari, Phillip Geyer, "Evaluierung der Darken-LBV-Gleichung zur Vorhersage von multikomponenten Diffusionskoeffizienten", research project, 2014.
- Sandra Haase, "Validierung eines mikrofluidischen Versuchsaufbaus für Diffusion in Flüssigkeiten mit konfokaler Ramanspektroskopie", master's thesis, 2015.
- Julia Thien, "Experimentelle Bestimmung des Flüssig-flüssig-Gleichgewichts und der Diffusionskoeffizienten eines ternären Systems mit konfokaler Raman-Spektroskopie", master's thesis, 2015.

- Andreas Hüttermann, "Ermittlung von Diffusionskoeffizienten in ternären Gemischen mittels 1D Ramanspektroskopie aus Batch-Versuchen", bachelor's thesis, 2015.
- Lasse Reinpold, Leonard Städtler, Maike Bruckhaus, "Mikrofluidische Messung der Diffusionskoeffizienten des Systems Toluol — Aceton – Wasser bei 25 °C mithilfe von konfokaler Raman-Spektroskopie", research project, 2017.

Kurzfassung

Diffusion ist limitierend in vielen (bio-)chemischen Prozessen. Daher sind Diffusionskoeffizienten nötig, um verfahrenstechnische Prozesse wie die Extraktion auszulegen. Diffusionskoeffizienten in flüssigen Mehrstoffgemischen sind jedoch häufig unbekannt, da Diffusionsexperimente zeitintensiv und aufwendig sind und Mehrstoffgemische mehrere Experimente zur Bestimmung eines Diffusionskoeffizienten benötigen.

Während die Mikrofluidik erlaubt, Experimentdauer und experimentellen Aufwand zu reduzieren, kann die Ramanspektroskopie die Anzahl notwendiger Experimente für Mehrstoffgemische senken. Um den Messaufwand für Diffusionskoeffizienten in Mehrstoffgemischen zu reduzieren, wurden in dieser Arbeit erstmals Mikrofluidik und Raman-Mikrospektroskopie zu Diffusionsmessungen in flüssigen Mehrstoffgemischen kombiniert: In einer mikrofluidischen H-Zelle fließen zwei Flüssigkeiten mit unterschiedlichen Konzentrationen parallel und lokale Änderungen durch Diffusion werden mit Raman-Mikrospektroskopie quantifiziert. Aus den Änderungen der Konzentration kann der Diffusionskoeffizient aus einem Diffusions-Konvektions-Modell bestimmt werden. So können Diffusionskoeffizienten in Mehrstoffgemischen aus einem Experiment bestimmt werden, während die bekannte Anfangskonzentration dank quantitativer Ramanspektroskopie zusätzliche Kalibriermessungen erübrigt. Die Experimentdauer liegt unter einer Stunde, der Probenverbrauch unter 6 mL.

Das Diffusionsmessverfahren wurde an binären und ternären Systemen validiert. Weiter wurden für binäre, ternäre und quaternäre Systeme neue Diffusionsdaten gemessen. Für ternäre Systeme reichen bereits zwei und für quaternäre Systeme drei Experimente zur Bestimmung des Diffusionskoeffizienten mit guter Genauigkeit. Für die Stoffsysteme zur Validierung des Verfahrens wurde eine gute Übereinstimmung mit der Literatur erzielt. Damit können Multikomponentendiffusionskoeffizienten schnell und mit geringem Aufwand für die Verfahrenstechnik bereitgestellt werden.

Abstract

Diffusion is the rate-limiting mass transport step in many (bio)chemical processes. Diffusion data is therefore necessary to design unit operations, *e.g.*, extraction. However, diffusion data on multicomponent mixtures in liquids are scarce, as diffusion measurements are intrinsically time-consuming and laborious. Also, several measurements are required for one diffusion coefficient matrix in a multicomponent mixture.

Microfluidics promises to reduce experiment time and experimental effort, while Raman spectroscopy reduces the number of necessary experiments for multicomponent mixtures. To reduce the measurement effort for diffusion coefficients in multicomponent mixtures, microfluidics and Raman microspectroscopy were combined in this work for the first time for the measurement of diffusion coefficient matrices in multicomponent mixtures: Two liquids of different concentrations co-flow in parallel in a microfluidic H-cell, while changes of concentration due to diffusion are quantified locally using Raman microspectroscopy. From these changes of concentration, the diffusion coefficient matrix is determined in a least squares procedure using a convection-diffusion-model. Thereby, diffusion coefficient matrices can be determined from a single experiment, while the known initial concentrations together with quantitative Raman spectroscopy spare additional calibration measurements. The experiment takes less than 1 h and the sample consumption is below 6 mL.

The developed method to determine diffusion coefficients was validated for binary and ternary systems. Additionally, new diffusion data was measured for a binary, a ternary and a quaternary system. Two and three experiments are sufficient to determine diffusion coefficient matrices with accuracy and precision for ternary and quaternary systems respectively. Systems used for validation agree with literature data. Thus, diffusion coefficient matrices of multicomponent mixture can be provided with small effort and in short time to enable the design of unit operations.

Chapter 1

Introduction

Diffusion is omnipresent in (bio)chemical processes, such as rectification and liquid-liquid extraction,[3] laminar mixing,[4–6] or generation of stable concentration gradients for cell-based studies in microfluidics.[7–9] While technical mixing processes proceed on the macroscale, diffusion is the dominant mixing process on the microscale. Since diffusion is often rate-limiting, knowledge of diffusion is fundamental for the design of microfluidic applications,[10–15] but also to understand industrial processes and environmental phenomena.[16,17]

However, diffusion data is scarce.[16,18–20] While the Dortmund Data Bank contains vapor-liquid equilibria on more than 9000 binary systems,[21] diffusion data is only available for little more than 2000 binary systems.[22] Moreover, the number of available diffusion data decreases one order of magnitude per additional component in the system. As technical systems contain more than two components in general, the lack of data on multicomponent systems is severe. Additionally, most multicomponent diffusion data is on aqueous mixtures only.[23]

If diffusion data is not available for a specific system, there are basically three options to generate data: molecular dynamics, engineering models, and experiments. Diffusion coefficients can be fully predictively computed from molecular dynamics (MD). This approach shows promising results for the Maxwell-Stefan (MS) and Fick diffusion coefficients if suitable force-field models are available.[24] Engineering models allow a rapid prediction of diffusion coefficients with moderate uncertainties.[25] Recent Darken-based models have a physically sound basis and improve prediction accuracies of Maxwell-Stefan diffusion coefficients in ideal multicomponent mixtures.[26]

Generally, additional data expressing thermodynamics is required to obtain the predictively relevant Fick diffusion coefficient from the Maxwell-Stefan diffusion coefficient.

Still, both MD and engineering models contain large uncertainties. Experimental multicomponent diffusion data are required if high accuracy is required. This data is also needed for validation and development of theoretical methods. Thus, experimental methods are required that provide accurate and precise values for diffusion coefficients in a short time and with little effort. To date, classical diffusion experiments allow for high accuracy and precision, but classical diffusion experiments are also time-consuming, laborious and require at least $n_c - 1$ experiments for a multicomponent system with n_c components.[27,28] In practice, multiple sets of $(n_c - 1)$ experiments are performed.[2,29]

Thus, an efficiently designed diffusion experiment is desirable. For this purpose, this work combines Raman spectroscopy and microfluidics: Raman spectroscopy provides independent concentration information on $n_c - 1$ species in a mixture from each measurement with high accuracy and high spatial resolution.[30,31] Consecutive, temporally and spatially resolved Raman measurements result in comprehensive concentration gradient information so that one diffusion experiment is sufficient to determine multicomponent diffusion coefficients.[32] Microfluidics enables short diffusion experiment time (<1 h), since the diffusion time is proportional to the squared diffusion distance.[33,34] Accordingly, binary diffusion coefficients have successfully been measured using microfluidics.[10,35–41] Also, it has repeatedly been hypothesized that microfluidics allows for the measurement of multicomponent diffusion coefficients.[5,31,36]

In this thesis, multicomponent diffusion coefficients have been measured using microfluidics for the first time.[17] For this purpose, microfluidics was combined with Raman spectroscopy to quantify all components in a multicomponent mixture simultaneously with high spatial resolution, resulting in a rapid and accurate measurement method for multicomponent diffusion coefficients. Diffusion coefficients are obtained from the measured local mole fractions using a convection-diffusion model. Diffusion coefficients were measured for four binary, three ternary and one quaternary systems. For three binary and two ternary systems, data was already available from literature and excellent agreement between the data was obtained. The novel diffusion coefficients measured in this work were compared to results from predictive engineering models. The presented method makes it possible to determine multicomponent diffusion coefficients from only one experiment. Low uncertainty in the diffusion coefficients requires only $n_c - 1$ experiments.

1.1 Structure of this Thesis

Chapter 2 gives an overview of the current state of the art for diffusion in liquids. This includes modeling diffusive transport and predictive engineering models used to determine diffusion coefficients. Classical diffusion measurements are presented and classified as benchmarks for new diffusion measurement methods. Characteristics of Raman spectroscopy and Raman spectroscopy for diffusion measurements are introduced. An introduction to microfluidic diffusion measurements completes chapter 2. The contribution of this thesis is deduced from the preceding discussion.

Chapter 3 presents details of multicomponent diffusion coefficient measurements using microfluidics and Raman microspectroscopy. The chapter starts with the characteristics of the measured systems and the experimental setup. Then, the convection-diffusion model required to determine diffusion coefficients is presented, followed by the model-based experimental procedure for accurate and precise diffusion coefficients. Data analysis and parameter estimation illustrates how diffusion coefficients are determined from experimental data.

Chapter 4 presents the results for the binary, ternary, and quaternary systems. The assumptions in the convection-diffusion model are discussed based on the experimental results. The determined diffusion coefficients are compared to literature and predictive engineering models.

Chapter 5 summarizes this thesis and conclusions are drawn regarding capabilities of multicomponent diffusion measurements using microfluidics and Raman microspectroscopy.

Chapter 2

State of the Art: Modeling and Measurement of Diffusion in Liquids

Parts of this chapter are adapted from Ref.[27] with permission from The Royal Society of Chemistry:

> C. Peters, L. Wolff, T. J. H. Vlugt, A. Bardow, "Diffusion in Liquids : Experiments, Molecular Dynamics, and Engineering Models". In: "Non-equilibrium thermodynamics with applications", ed. by Dick Bedeaux, Signe Kjelstrup, Jan Sengers, 2016, p. 78-104.
> `https://pubs.rsc.org/en/content/chapter/bk9781782620242-00078/978-1-78262-024-2`

Contribution report: Writing the draft of 5.1, 5.2, 5.4 and 5.5 together with Ludger Wolff, principal author together with Ludger Wolff.

Fig. 2.2 is reprinted with permission from

> C. Peters, J. Thien, L. Wolff, H.-J. Koß, A. Bardow (2019). Quaternary Diffusion Coefficients in Liquids from Microfluidics and Raman Microspectroscopy: Cyclohexane + Toluene + Acetone + Methanol. *J. Chem. Eng. Data*, `https://doi.org/10.1021/acs.jced.9b00632`.[42] Copyright (2019) American Chemical Society.

Contribution report: Writing the draft, principal author, planning the experimental setup, choosing the chemical systems, planning the experiments, supporting experiments conducted by J. Thien and R. Becka, data evaluation.

This Chapter provides an overview of diffusion in liquids covering its modeling, prediction and measurement. In section 2.1, the phenomenon of diffusion is introduced, while the modeling of diffusive transport is presented in section 2.2. Engineering models to predict diffusion coefficients in infinite dilution and concentrated solutions are introduced in section 2.3. Section 2.4 presents and characterizes classical diffusion measurement methods as benchmark for new diffusion measurement methods. In section 2.5, characteristics and limitations of Raman spectroscopy are discussed together with the advantages and disadvantages of its use for diffusion measurements. Possibilities of Raman microspectroscopy and microfluidics for diffusion measurements are highlighted in section 2.6. The contribution of this thesis is deduced from the preceding discussion in section 2.7.

2.1 Diffusive Mass Transport

Mass transfer plays an important role in all kinds of processes in (bio)chemistry. Mass transfer is often limited by diffusion, especially on the microscale. As diffusion is often a rate-limiting process, diffusion coefficients are required for the design of industrial and microfluidic processes as well as the modeling of environmental phenomena.[3,13,43] Diffusion was first described already in 1855 by Adolf Fick.[44] Still, the prediction, modeling and measurement of diffusion coefficients are an ongoing field of research, especially in liquids.[45–49]

Mass transport is the net movement of mass between different locations. Commonly, this movement is divided into convection and diffusion. Convection refers to the movement of the mixture as a whole, where each species has the same velocity. Diffusion is the movement of molecules in the mixture relative to the velocity of the fluid, which is called the reference velocity of this fluid. This reference velocity can be a mass or molar averaged reference velocity. The diffusion flux depends on the chosen velocity reference frame, *i.e.*, a molar diffusion flux accompanies a molar averaged reference velocity. Taylor and Krishna[3] give a detailed overview on reference frames and their suitability for different applications. In this work, the molar (M) and volume (V) average reference velocity frames are used. The volume average reference velocity is used to describe diffusion measurements, while diffusion coefficients are predicted from correlations in the molar average reference velocity frame.

Different types of diffusion are distinguished, of which mutual diffusion and self-diffusion are of practical importance. Mutual diffusion, also called inter-diffusion, denotes the movement of species i relative to the reference velocity as shown in Fig. 2.1A). Mutual diffusion occurs due to chemical-potential gradients such as induced by concentration or temperature differences.[50] Self-diffusion, also called intra-diffusion, refers to movements of species i due to Brownian motion and describes local concentration fluctuations of i as shown in Fig. 2.1B). In binary mixtures, self- and mutual diffusion coefficients are the same for the limit of infinite dilution ($x_i \rightarrow 0$). As the topic of this work is mutual diffusion, diffusion denotes mutual diffusion unless otherwise noted.

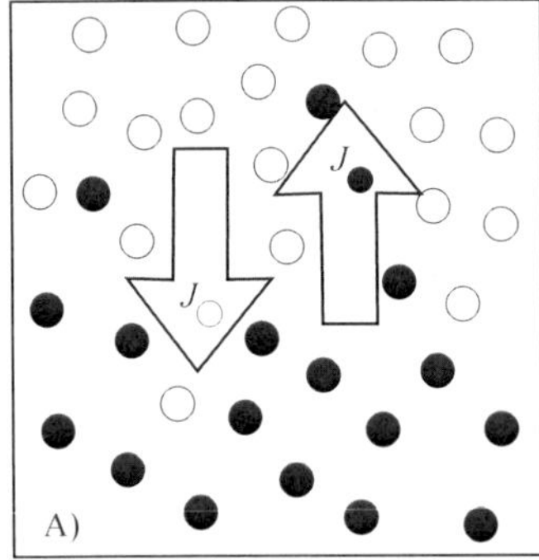

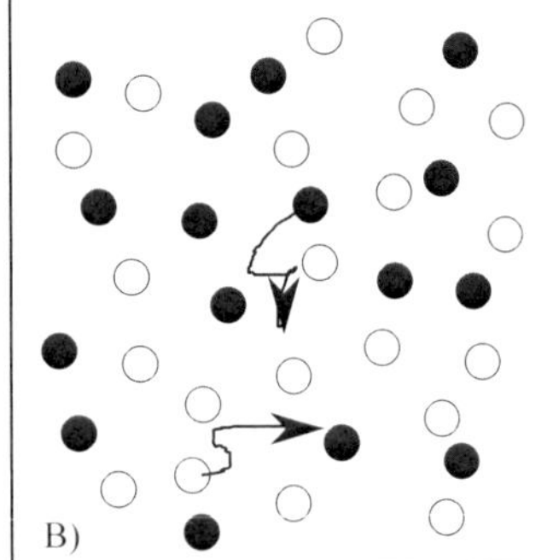

Figure 2.1: A) Mutual or inter-diffusion over concentration gradient resulting in the net diffusive fluxes $J_\bullet$ and $J_\circ$. B) Self or intra-diffusion (local fluctuations without net flux).

The diffusion coefficient is the constant coupling of the driving potential for diffusion and the diffusive flux (section 2.2). For liquids, the diffusion coefficient D is $\mathcal{O}(1 \times 10^{-9}\,\mathrm{m\,s^{-2}})$. The diffusion coefficient depends on the composition of the mixture x_i, its temperature T, and pressure p. The temperature-dependence is in the range of $\sim 1\,\%\,\mathrm{K}^{-1}$,[25] while the pressure-dependence is less pronounced with $\sim 1\,\%\,\mathrm{MPa}^{-1}$.[25,51] The form of the concentration-dependence varies strongly with the type of system, *i.e.*, whether the system is non-ideal or ideal.

To date, no general valid model structure for the concentration-, temperature-

and pressure-dependence is known. Thus, a variety of models for $D(p,T,x)$ have been proposed.[3] Experimental data are required for the discrimination of different engineering models to predict diffusion coefficients as well as for the parametrization of these models.[46,52] For this purpose and also for the modeling of processes, reliable and accurate measurements of diffusion coefficients are needed.

2.2 Modeling Diffusive Transport

In this section, the mathematical description of multicomponent mutual diffusion is presented using the generalized Fick's law (section 2.2.1) and the Maxwell-Stefan approach (section 2.2.2).

2.2.1 Generalized Fick's Law

The generalized Fick's law relates the diffusive fluxes to the diffusion coefficients and the concentration gradients. In the molar reference frame M, generalized Fick's law reads

$$J_i^{\mathrm{M}} = -c_{\mathrm{t}} \sum_{j=1}^{n_c-1} D_{ij}^{\mathrm{M}} \nabla x_j \,, \quad i = 1, \ldots n_c - 1 \tag{2.1}$$

with the molar diffusive flux J_i^{M}, the spatial gradient ∇ of the mole fraction x_j as driving force and the total molar concentration c_{t}. n_c denotes the number of components in the system and D_{ij}^{M} are the entries of the Fick diffusion coefficient in the molar reference frame.

In the volume reference frame V, the system is described by molar diffusion fluxes J_i^V relative to volume average velocity. Here, the generalized Fick's law is

$$J_i^V = -\sum_{j=1}^{n_c-1} D_{ij}^V \nabla c_j \,, \quad i = 1, \ldots n_c - 1 \tag{2.2}$$

with the entries of the Fick diffusion coefficient in the volume reference frame D_{ij}^V and the concentration gradient ∇c_j as the driving force. Here, c_j denotes the molar concentration of component j.

The main diffusion coefficient D_{ii} is the proportionality factor for the flux of species i due to its own concentration gradient. The cross-diffusion coefficient D_{ij} quantifies the flux of i due to a concentration gradient in j. The diffusion coefficient matrix $\boldsymbol{D}$ must be positive definite, *i.e.*, all eigenvalues, marked by †, are positive[3]

$$D_i^{\dagger} > 0 \ . \tag{2.3}$$

The main diffusion coefficients D_{ii} often dominate the diffusive flux, as often the main diffusion coefficients are larger than the cross-diffusion coefficients $D_{ii} > \|D_{ij}\|$.[3] Also, the main diffusion coefficients D_{ii} are usually positive, while the cross-diffusion coefficients D_{ij} can be of either sign, depending on the non-ideality of the system.[3,53] In rare cases, individual main diffusion coefficients D_{ii} can have a negative sign, as long as the diffusion coefficient matrix $\boldsymbol{D}$ remains positive definite,[53] *i.e.*, $D_i^{\dagger} > 0$, including a positive trace $\mathrm{tr}(\boldsymbol{D}) > 0$ of the matrix, with

$$\mathrm{tr}(\boldsymbol{D}) = \sum_i D_{ii} = \sum_i D_i^{\dagger} \ . \tag{2.4}$$

Cross or negative main diffusion coefficients can lead to uphill diffusion in multi-component mixtures.[54] Uphill diffusion denotes diffusion of a component against its concentration gradient.

The diffusion coefficient matrix $\boldsymbol{D}$ can be converted between the molar M and the volume V reference frames according to[3]

$$\boldsymbol{D}^V = \boldsymbol{B}^{Vu} \boldsymbol{D}^{\mathrm{M}} \boldsymbol{B}^{uV} \ , \tag{2.5}$$

with

$$B_{ij}^{Vu} = \delta_{ij} - x_i(\bar{V}_j - \bar{V}_{n_c})/\bar{V}_t \quad \text{and} \tag{2.6}$$

$$B_{ij}^{uV} = \delta_{ij} - x_i(1 - \bar{V}_j)/\bar{V}_{n_c} \ . \tag{2.7}$$

Here, δ_{ij} denotes the Kronecker delta, $\bar{V}_t$ the mixture molar volume, $\bar{V}_i$ and $\bar{V}_{n_c}$ the partial molar volume of component i and n_c, respectively.

2.2.2 Maxwell-Stefan Equations

The Maxwell-Stefan equations are closely linked to the thermodynamics of irreversible processes.[3,55,56] The Maxwell-Stefan approach specifies that the chemical-potential gradient is balanced by the friction forces between the moving molecules

$$-\frac{1}{RT}\nabla\mu_i|_{T,p} = \sum_{j=1,j\neq i}^{n_c-1} \frac{x_j\left(\boldsymbol{\nu}_i - \boldsymbol{\nu}_j\right)}{\text{Đ}_{ij}} , \tag{2.8}$$

with the gas constant R, the absolute temperature T, and the Maxwell-Stefan diffusion coefficients as inverse friction coefficients Đ_{ij}. The gradient of the chemical potential $\nabla\mu_i$ of component i at constant p, T, x_i is balanced by friction forces due to the difference $(\boldsymbol{v}_i - \boldsymbol{v}_j)$ between the average velocities of components i and j. The Maxwell-Stefan diffusion coefficients are symmetric

$$\text{Đ}_{ij} = \text{Đ}_{ji} \tag{2.9}$$

according to the Onsager reciprocal relations.

The Maxwell-Stefan description of diffusion can be written in a similar form as the generalized Fick's law in the molar reference frame M solved for fluxes:[55,57]

$$\boldsymbol{J}^{\mathrm{M}} = c_{\mathrm{t}}\boldsymbol{B}^{-1}\boldsymbol{\Gamma}\nabla\boldsymbol{x} \tag{2.10}$$

with the vector $\boldsymbol{J}^{\mathrm{M}}$ of the diffusive fluxes J_i^{M}, the vector $\boldsymbol{x}$ of the mole fractions x_i, and the inverse of the Maxwell-Stefan diffusion coefficient matrix $\boldsymbol{B}$ with the elements

$$B_{ii} = \frac{x_i}{\text{Đ}_{in_c}} + \sum_{k=1,i\neq k}^{n_c} \frac{x_k}{\text{Đ}_{ik}} \quad \text{and} \tag{2.11}$$

$$B_{ij} = -x_i\left(\frac{1}{\text{Đ}_{ij}} - \frac{1}{\text{Đ}_{in_c}}\right) . \tag{2.12}$$

$\boldsymbol{\Gamma}$ denotes the thermodynamic factor which describes the deviation from ideal mixing behavior with

$$\Gamma_{ij} = \delta_{ij} + x_i \left(\frac{\partial \ln \gamma_i}{\partial x_j} \right)_{T,p,\sum} \tag{2.13}$$

and the activity coefficient γ_i of component i. For the derivative of the activity coefficient at constant temperature, pressure and mole fractions, the mole fraction x_j is balanced by the n_c-th so that the constraint $\sum_{i=1}^{n_c} x_i = 1$ is satisfied (indicated by the symbol $\sum$).[58,59] The thermodynamic factor $\boldsymbol{\Gamma}$ is not symmetric, but its determinant must be positive.[53]

As can be seen from the generalized Fick's law (2.1) and the Maxwell-Stefan formulation (2.10), the relation between the Maxwell-Stefan and Fick diffusion coefficient is[3]

$$\boldsymbol{D}^{\mathrm{M}} = \boldsymbol{B}^{-1}\boldsymbol{\Gamma} \, . \tag{2.14}$$

Both diffusion coefficient matrices, $\boldsymbol{D}$ and $Đ$, are positive definite. Due to the symmetry, there are only $n_c(n_c - 1)/2$ Maxwell-Stefan diffusion coefficients, but there are $(n_c - 1)^2$ Fick diffusion coefficients. Therefore, the entries of the Fick diffusion coefficient matrix are not independent for multicomponent systems ($n_c > 2$). Also, Fick diffusion coefficients are strongly concentration-dependent, as the thermodynamic factor $\boldsymbol{\Gamma}$ is also concentration dependent.[3]

The thermodynamic factor is not directly accessible in experiments as the gradient of the chemical potential $\nabla\mu_i$ is not directly measurable. Also, the calculation of the thermodynamic factor with G^{E}-models or equations of state (EOS) introduces large uncertainties.[3,58,60] The uncertainties are inevitable as G^{E}-models and EOS were not designed for diffusion, but phase equilibria. Also, G^{E} and EOS parameters are generally fitted to phase equilibrium data and not diffusion data.[61] Recent approaches to calculate the thermodynamic factor from MD seem promising,[62,63] but are not established yet and rely on suitable force fields. Thus, the application of the Maxwell-Stefan equations is limited in experiments, but dominant in modeling. Fick's generalized law is more accessible in practical applications because only concentrations and no chemical potentials are required.

The reliability of modeling diffusive transport, independent of the Fick and the Maxwell-Stefan approach, depends on the quality of available diffusion coefficients. In general, diffusion coefficients can be obtained from experiments (see Sec. 2.4), engineering models (see 2.3) or MD. Using MD simulations, Fick diffusion coefficients or Maxwell-Stefan diffusion coefficients are accessible.[24,64] Recently, very promising results (0 % to 20 %) were obtained for ideal to complex mixing behavior,[24,64] while strongly non-ideal mixtures showed deviations of 15 % to 20 %.[64] Still, MD simulations are time-consuming[65] and to assess recent developments of MD and force fields for the determination of diffusion coefficients, experimental data is needed.

2.3 Prediction of Multicomponent Diffusion Coefficients in Concentrated Solutions

Engineering models are widely used to predict diffusion coefficients, *e.g.*, in infinite dilution, in concentrated solutions or regarding the temperature-dependence of the diffusion coefficient.[26,27,50,66] These engineering models use physical properties such as molar volume V or viscosity η of the system for the prediction. The advantage of these models is that only little data is required. Still, depending on the required conditions and physical data to start with, several steps are necessary to calculate a diffusion coefficient in a concentrated solution. If a diffusion coefficient $D_{\mathrm{AB}}^{x_{\mathrm{A}}\to 0}$ of component A at infinite dilution in B is missing, it has to be predicted. For this purpose, the Stokes-Einstein equation[50]

$$D_{\mathrm{AB}}^{x_{\mathrm{A}}\to 0} = \frac{k_B T}{6\pi \eta_{\mathrm{B}} r_{\mathrm{A}}} \tag{2.15}$$

can be used, with the Boltzmann constant k_{B}, the molecular radius of solute A r_{A} and the viscosity of the solvent η_{B}. These predictions have an accuracy of 70 % regarding the experimental value,[67] which can be increased to 90 % using the more popular Wilke-Chang equation

$$D_{\mathrm{AB}}^{x_{\mathrm{A}}\to 0} = 7.4 \times 10^{-8} \frac{(\phi_{\mathrm{B}} M_{\mathrm{B}})^{1/2} T}{\eta_{\mathrm{B}} V_{\mathrm{A}}^{0.6}} , \tag{2.16}$$

or modifications of this model.[25,50] Here, M_{B} denotes the molecular weight of solvent B and ϕ_{B} the association factor of solvent B ranging from 2.6 for water to 1.0 for unassociated solvents.[50]

To determine a Maxwell-Stefan diffusion coefficient in concentrated solutions, further predictive engineering models are necessary, which need diffusion coefficients at infinite dilution or self-diffusion coefficients D_{self}. For binary mixtures, such a model is the Darken equation[68]

$$Đ_{ij} = x_i D_{j,\mathrm{self}} + x_j D_{i,\mathrm{self}} \ . \tag{2.17}$$

It is originally an empirical mixing rule but can also be derived from statistical-mechanical theory in a limiting case.[26] It was also shown to be superior to other models in recent molecular simulation studies. For multicomponent systems, Liu, Bardow, and Vlugt[26] suggested the predictive Darken-LBV equation. One part of this equation is the multicomponent Darken equation

$$Đ_{ij} = \frac{D_{i,\mathrm{self}} D_{j,\mathrm{self}}}{D_{\mathrm{mix}}} \quad \text{with} \quad D_{\mathrm{mix}} = \sum_{i=1}^{n} \frac{x_i}{D_{i,\mathrm{self}}} \ , \tag{2.18}$$

where D_{mix} denotes the effective self-diffusion coefficient of the solution. The multicomponent Darken equation was derived from linear response theory and the Onsager relations. The second part of the predictive Darken-LBV equation is a predictive model for the self-diffusivities

$$\frac{1}{D_{i,\mathrm{self}}} = \sum_{j=1}^{n} \frac{x_j}{D_{i,\mathrm{self}}^{x_j \to 1}} \ . \tag{2.19}$$

The predictive Darken-LBV equation performed well (average deviation of 8 % to 11 %) for ternary and quaternary WCA systems as well as n-hexane + cyclohexane + toluene.[69]

A survey on further predictive models for Maxwell-Stefan diffusion coefficients in concentrated binary and multicomponent solutions is given in Appendix A. Generally, the Maxwell-Stefan diffusion coefficient and the self-diffusivities coincide at infinite dilution

$$D_{i,\mathrm{self}}^{x_j \to 1} = Đ_{ij}^{x_j \to 1} \; . \tag{2.20}$$

as described by the Darken equation (2.17). Maxwell-Stefan and Fick diffusion coefficients also coincide at infinite dilution, as the thermodynamic factor (eq. (2.13)) converges with the identity matrix:

$$D_{ij}^{x_j \to 1} = Đ_{ij}^{x_j \to 1} \; . \tag{2.21}$$

The thermodynamic factor $\boldsymbol{\Gamma}$ (2.13) to calculate Fick from Maxwell-Stefan diffusion coefficients can be obtained using activity coefficients from, *e.g.*, G^{E}-models as NRTL and COSMO-RS or an equation of state (EOS) as PC-SAFT.

If diffusion coefficients are unavailable at the chosen temperature, a few correlations for the temperature-dependence are listed in Appendix A.2.[50] A straight-forward model,

$$\frac{D_{\mathrm{AB}}^{x_\mathrm{A} \to 0} \eta_\mathrm{B}}{T} = \mathrm{constant} \; , \tag{2.22}$$

is based on the Wilke-Chang eq. (2.16) but its use should be restricted to small temperature ranges (30 K).[70]

Currently, predicted diffusion coefficients still contain uncertainties of 10 % or more, from the model itself and, if applied, the thermodynamic factor. Thus, models have not yet achieved satisfactory accuracy.[25] The long-term expectation is that multicomponent diffusion coefficient matrices $\boldsymbol{D}$ can be predicted in simple fluids with high accuracy as the models and the quality of the underlying (binary) data improve. New and refined models still have to be tested on experimental data. Thus, experimental data remains indispensable.

2.4 Classical Diffusion Experiments

Experiments are still the only way to obtain diffusion coefficients of high precision and accuracy, both in binary and multicomponent liquid systems. As Fick diffusion coefficients are directly accessible in experiments in comparison to Maxwell-Stefan diffusion coefficients requiring the thermodynamic factor, Fick diffusion coefficients are typically measured. Diffusion experiments are time-consuming, laborious, expensive, and difficult in particular for multicomponent systems.[28,71] To determine the concentration-dependence of the diffusion coefficient, experiments at different concentrations are required.[72] Only with temporally and spatially resolved concentration measurements as well as knowledge of the excess volume, the diffusion coefficient can be determined for all concentrations in one experiment.[73]

For one multicomponent diffusion coefficient, the required experimental information is $\propto (n_c - 1)^2$. For a hypothetical experiment with only one measurement of a scalar information, *e.g.*, one refractive index, $(n_c - 1)^2$ independent experiments would be necessary.[29] Here, independent measurements are at different times or positions within one experiment or in experiments with different initial concentrations, *i.e.*, independent experiments.

Classical experiments consist of several measurements ($\geq n_c - 1$) of scalar information, *e.g.*, of the refractive index, but the scalar information does not allow for the determination of $n_c - 1$ temporally and/or spatially resolved concentrations.[27,29] Therefore, $(n_c - 1)$ independent experiments are required. For ternary diffusion coefficient matrices, a minimum number of two experiments is required, except for 1D Raman measurements, see section 2.5. Additionally, for reliable results, at least twice the minimal number of experiments is recommended.[29]

While the number of experiments increases with every component, the number of available diffusion coefficients decreases; the more components are in a system, the less systems are available from literature as shown in Fig. 2.2.

To introduce diffusion experiments, table 2.1 gives an overview on established and recent measurement methods. The table content is adapted from Peters *et al.*[27] and based on Woolf *et al.*[28] The table specifies measurement conditions, such as temperatures, pressures and experiment duration. Whenever using experimental data, attention should also be payed to the experimental details and the quality of data, as experimental data always has a measurement uncertainty depending on the techniques used. Typical deviations between different publications of diffusion coefficients can be found, *e.g.*, for cyclohexane + toluene.[75–78] Therefore, the overview on the diffusion measurement methods also includes the accuracies. Generally, tem-

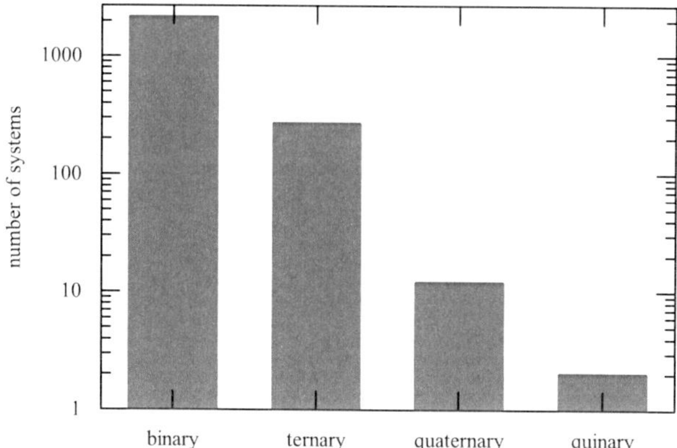

Figure 2.2: Number of binary and multicomponent systems for which diffusion coefficients are published.[22,74]

perature control is crucial for diffusion measurements as diffusion coefficients are strongly temperature-dependent, while the pressure is a minor error source (see also Sec. 2.1).[25]

A detailed overview on classical diffusion experiments was given by Woolf *et al.*,[28] short descriptions were also compiled by Winkelmann.[79]

In the diaphragm measurement technique for diffusion coefficients, two solutions of equal volumes but difference in concentrations are separated by a porous membrane. The concentration difference between the solutions should be preferably small as an effective diffusion coefficient is measured. The concentration difference needs to be larger than the sensitivity of the used analytic technique.[28] The solutions are stirred such that the concentrations in each solution are homogenous. The concentrations in the solution change over time due to diffusion over the membrane. The change of the concentrations is measured over time. Then, the diffusion coefficient is determined from the temporally resolved concentrations. Disadvantages of the diaphragm measurement technique are necessary calibration measurements to deter-

Table 2.1: Measurement techniques for mutual and self-diffusion coefficients, adapted from Peters *et al.*[17] and based on Woolf *et al.*[28]

	Accuracy in %	Temperature range in K	Pressure range in MPa	Experiment duration	Min. no. of experiments for ternary system
Diaphragm cell	0.5 to 1	< 400	≤ 400	> 2 d	2
Conductance	0.2	≈ 298	≈ 0.1	> 2 d	2
Taylor dispersion	≈ 1	≤ 600	≤ 100	< 1 h	2
NMR	≈ 1 to 2	≤ 700	≤ 100	< 1 h	-
Gouy	< 0.1	≈ 298	≈ 0.1	(1 to 2) d	2
Rayleigh	< 0.1	≈ 298	≈ 0.1	(1 to 2) d	2
DLS	≈ 0.8 to 10[80]	283 to 323[80]	≈ 0.1	≈ 1 h[81]	[a]
Raman	≈ 0.2 to 5[70,82]	≤ 350[70]	≤ 40[83]	< 1 h	1[32]
Microfluidic	≈ 1 to 2[36]	≤ 400[76]	≈ 0.1	< 1 h	1 in situ[b]

[a] Only eigenvalues of Fick diffusion matrix[84]
[b] As shown in section 4

mine the membrane characteristics, which are specific for every chemical system to be measured. Overall, the diaphragm measurement technique is multifunctional as it can be used for tracer, self- and mutual diffusion. Diffusion coefficients have been published for systems of two, three, and more components.[85] Over the last years, the diaphragm measurement technique has been widely replaced by Taylor dispersion and interferometry.[23]

The Taylor dispersion needs no system-dependent calibration and reduces the experiment duration significantly (Table 2.1) while increasing the measurement uncertainty slightly to 1 %. Nowadays, Taylor dispersion is widely used as it profits from a simple experimental setup, *e.g.*, employing conventional high pressure liquid chromatography (HPLC) equipment.[86] Short measurement duration, wide operating conditions, comparatively low sample consumption and a physically sound model of the mass transport complete the advantages.[87] Taylor dispersion is based on injecting a sample of different concentration into the laminar flow of the solution. For this purpose, often an unpacked, coiled column is used as capillary in an HPLC oven, which allows easy temperature control. Simultaneously, the HPLC detection unit can be used to measure time-resolved concentrations at the outlet. The concentration in radial direction is almost homogenous as the diffusion in radial direction is complete within the retention time. Due to the parabolic velocity profile, the sample is widened and deformed to a time-dependent Gaussian profile. From the retention time at column outlet and the time-dependent concentration, the diffusion coefficient can be determined. Typically, the refractive index is measured by the detection unit. The most important disadvantage for the measurement of multicomponent diffusion coefficients is this detection unit: For a multicomponent diffusion coefficient matrix, at least $n_c - 1$ independent experiments are necessary as the refractive index gives only 1 information per measurement. A minor disadvantage is the high uncertainty compared to diaphragm cell and interferometric methods. Overall, the largest advantages are the short experiment time and the availability of the equipment. The model of the Taylor dispersion has been extended to multicomponent mixtures so that ternary[86,88,89], quaternary[90,91], and quinary[74] diffusion coefficient matrices were already measured with Taylor-dispersion.

The Gouy interferometric method uses a vertical diffusion cell, in which two homogenous liquids of different concentration are layered on top of each other. Monochromatic light transverses the diffusion cell perpendicular to the diffusion at different heights.[28,92] The resulting fringe patterns are recorded behind the diffusion cell and converted into spatially resolved refractive indices. For binary systems, temporally and spatially resolved concentrations can be directly determined from the refractive index and consequently the diffusion coefficient. Multicomponent diffusion coefficient matrices can also be calculated from the parameters describing the fringe pattern. The Rayleigh interferometric method is similar, but the monochromatic light is also directed to a reference cell in addition to the diffusion cell.[28] Adversely, both interferometric methods require at least $n_c - 1$ independent measurements for multicomponent systems and long experiments.[27,93] However, if low uncertainties are required, Gouy and Rayleigh interferometric methods provide the highest accuracy (Table 2.1).

Dynamic light scattering (DLS) enables fast measurements for mutual diffusion coefficients, the experiment time is comparable to Taylor dispersion. Another advantage is that measurements can be conducted in homogeneous samples as the scattering arises from microscopic fluctuations in the sample. A laser is directed on a probe volume and the scattered light is detected under an angle, *e.g.*, 90°.[80,94,95] To determine the diffusion coefficient, an autocorrelation function is fitted to the time-dependent intensity of the scattered light. The diffusion coefficient is one of the parameters of the autocorrelation function, together with the scattering vector in the decay rate. DLS is particularly suitable for macromolecules due to their good scattering properties.[96] However, the uncertainties vary depending on the system and reach up to 10 %.[80] In contrast, the experiment duration is competitive to Taylor dispersion with below 1 h.[81] For multicomponent system, only the eigenvalues of the diffusion coefficient matrix are accessible.[84,97] Thus, DLS is interesting to measure multicomponent diffusion coefficients if the cross-diffusion coefficients are negligible. As many systems are non-ideal, the missing cross-diffusion coefficients limit the practical interest of DLS for multicomponent systems.

Regarding advantages and disadvantages of classical diffusion experiments, no method has yet shown a combination of all advantages: short experiment duration, no calibration of the setup, low uncertainties and a full multicomponent diffusion coefficient matrix from less than $n_c - 1$ experiments.

2.5 1D Raman Spectroscopy for Diffusion Measurements

Raman spectroscopy is interesting for diffusion measurements because Raman measurements are *in situ* and non-invasive, like refractive-index-based techniques. However, Raman spectroscopy quantifies all components simultaneously. The accuracy of mole fractions is typically 0.1 mol % to 1 mol %.[82,98,99] Additionally, high spatial resolutions can be achieved. The theoretical diffraction limit in Raman microspectroscopy is $0.61\lambda/\mathrm{NA}$ with the numerical aperture NA and the wavelength λ. In practical applications, a spatial resolution of $\mathcal{O}(1\,\mu\mathrm{m})$ and a depth resolution of $\mathcal{O}(5\,\mu\mathrm{m})$ are feasible for a wavelength $\lambda = 532\,\mathrm{nm}$.[100,101] The possibility of 1D Raman spectra has already been highlighted in the 1970s.[102] Thus, Raman spectroscopy can be used to measure temporally and spatially resolved mole fraction profiles for the determination of diffusion coefficients. In the following, the Raman effect and the special features of diffusion measurements using Raman spectroscopy are discussed.

2.5.1 Raman Spectroscopy

Raman spectroscopy is based on the Raman effect, which was discovered in 1928 by C. V. Raman.[103] The spontaneous Raman effect describes inelastic light scattering caused by light stimulation due to electrons switching from one vibrational state to another vibrational state passing a virtual state, see Fig. 2.3. The Raman effect concerns only 10^{-5} to 10^{-7} of the scattered electrons.[104] The majority of scattered light is scattered elastically, known as Rayleigh scattering.

The Raman scattering is divided into Stokes and anti-Stokes scattering. While the Stokes scattering denotes energy transfer from a photon to a molecule, the anti-Stokes scattering describes an energy transfer from a molecule to a photon, provided the electron was on an increased vibrational state. As the temperature influences the number of electrons on increased vibrational states, the ratio of Stokes to anti-Stokes scattering depends on the temperature. At room temperature, Stokes scattering is dominant because the electrons dominantly occupy the low vibrational states. For Stokes scattering, an emitted photon is shifted to longer wavelengths as its energy decreases.

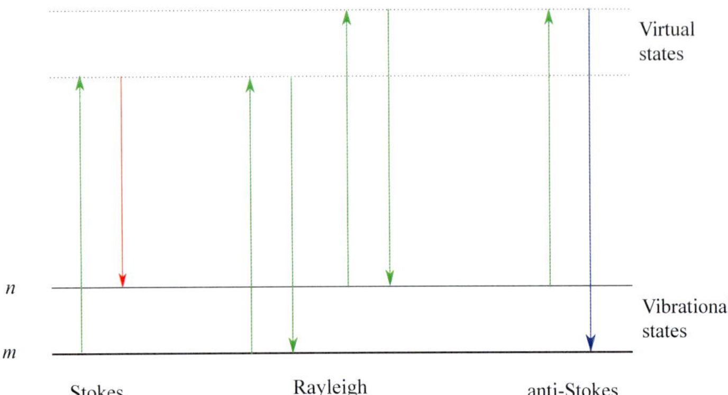

Figure 2.3: Raman and Rayleigh scattering process based on vibrational and virtual states.

The Raman effect is applicable if the components are Raman active, *i.e.*, the electron shell can be polarized. The Raman effect is observable in the gaseous, liquid, and solid state.

For visualization, the scattering is recorded in a spectrum *i.e.*, spectrally resolved photon counts. A Raman spectrum provides various bands at spectral positions, which are characteristic for every component and its state of matter. For liquid toluene, the spectral bands are shown in Fig. 2.4. A spectrum shows the intensity, *i.e.*, the number of scattered electrons, as a function of the relative wavenumber ν. Toluene provides distinctive bands in the fingerprint region ($500\,\mathrm{cm}^{-1}$ to $1500\,\mathrm{cm}^{-1}$), which include aromatic rings, CH_2, CH_3, and generally aliphatic chains. Additionally, toluene shows an OH-band from $2900\,\mathrm{cm}^{-1}$ to $3500\,\mathrm{cm}^{-1}$ and a methyl group-band at $2900\,\mathrm{cm}^{-1}$.

Raman spectra of mixtures often show nonlinear effects in comparison to the spectra of the pure components, *e.g.*, a spectral band shift or a change of the band's shape in a mixture. The reason for this is deformation of the electron shell due to molecular interactions. Typical examples are the change of the water OH-band with temperature or in the presence of salts.[105–107]

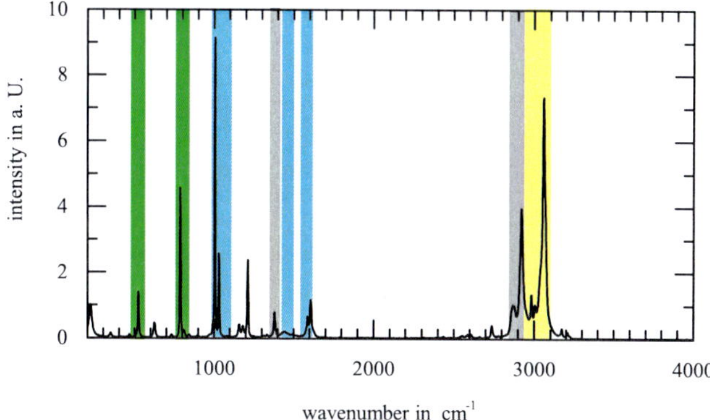

Figure 2.4: Raman spectrum of liquid toluene showing spectral bands of mono-substituted benzene (■), aromatic ring (■), methyl group (■), and aromatic CH (■).

Setups for spontaneous Raman spectroscopy are often based on compact commercial devices or tailor-made in academia. The setup generally contains a laser as monochromatic light source, a probe volume and a spectrometer. The scattering is observed either in direction of the laser beam, perpendicular to the laser beam or in backscattering. The setup can include a dichroic mirror to avoid Rayleigh scattering in the spectrometer in case of forward or backscattering. Central elements of the spectrometer are a prism or a grating to decompose the Raman scattering into different wavelengths, and the detection unit, *e.g.*, a charge coupled device (CCD) detector. The spectrometer can contain a narrow slit on which the sample volume is imaged to reduce the sample volume. Thus, the origin of the light behind the slit is restricted to the sample volume. A notch filter or a band-pass filter is used to suppress the Rayleigh scattering. Depending on the scattering cross-section of the observed components, experimental parameters such as laser intensity, saturation of the detector, wavenumber range, repeated acquisitions and signal to noise ratio should be chosen.

Raman spectroscopy allows different configurations. Each spectrum corresponds to the complete sample volume, which is imaged on the detection device.[108] For 1D spectra, one dimension of the detector is used for one axis of the sample, while the other dimension of the detector records the spectral shifts.[102] To acquire 2D spectra, the Raman scattering of one sample volume is projected on several defined areas of the detector.[109,110] Each area collects spectrally integrated intensities and the columns and lines of the detector areas can be converted to a spatial 2D resolution. As many spectral information are recorded as areas are defined on the detector.

Overall, Raman spectroscopy can be used to identify components using libraries, for structure determination and for the simultaneous quantification of all components in mixtures based on distinctive bands and the Lambert-Beer law stating a relation between signal and substance amount.[30,99,111,112] For liquids, only moderate exposure times are necessary (~0.1 s to 10 s).

2.5.2 Diffusion Measurements Using 1D Raman Spectroscopy

Raman spectroscopy is particularly suitable for diffusion measurements as shown by Bardow *et al.*[82] In a batch experiment, two solutions containing all components are layered on top of each other. The diffusion process is then observed *in situ* with 1D Raman spectroscopy. A scheme of the setup is shown in Figure 2.5.

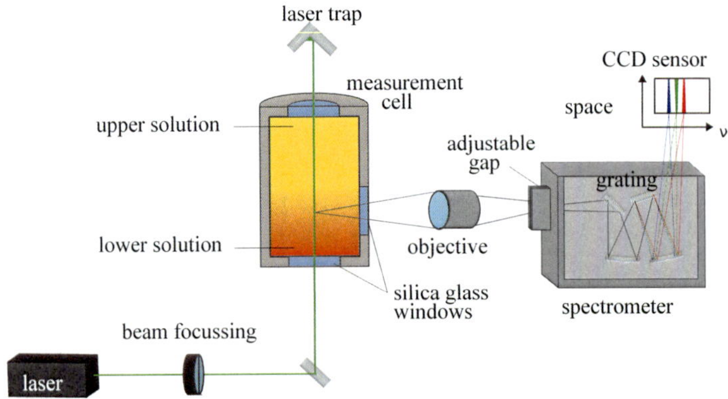

Figure 2.5: Schematic setup of 1D Raman spectroscopy diffusion measurement.

During the diffusion experiment, a laser beam traverses the diffusion cell in the direction of diffusion. The Raman scattering is observed spatially resolved along the observation line perpendicular to the direction of diffusion with a spectrometer.[82] The observation line starts at the bottom of the diffusion cell, traverses the interface and ends in the second bulk phase. The CCD-chip of the spectrometer collects the spatially resolved 1D Raman spectra over time. A spatial resolution of approximately 100 µm is achieved. The theoretical spatial resolution is 25 µm but due to imperfections in the optical system, one point is imaged on 4 lines of the CCD chip.

These 1D spectra are converted to spatially and temporally resolved mole fraction profiles of all components as shown in Figure 2.6. The large number of spatially resolved mole fraction profiles results in a high information content per experiment, which allows even to determine the concentration-dependence of the diffusion coefficient from a single experiment.[73]

In the estimation of the diffusion coefficient, measured mole fraction profiles are compared to a 1D diffusive mass transport model. The model describes diffusion in a restricted volume, including no mass fluxes over the lower and upper boundary. The observed zone in the diffusion cell includes the bottom and the inflection point. The upper boundary is outside the observed region ($l_{max} \geq 10$ mm). The diffusion coefficient can be determined from the mass transport model either analytically assuming an initial step in the concentration[113] or numerically with the first concentration profile as initial condition.[70]

Model-based experimental design methods have revealed that the maximal information on the diffusion coefficient is available as the diffusion reaches the wall of the diffusion cell.[82] Thus, it is recommended to observe the bottom of the diffusion cell. Hence, the experiment time is approximately 1 h for diffusion distances of approximately 10 mm depending on the diffusion coefficient. Additionally, model-based experimental design methods suggest for multicomponent diffusion experiments that initial concentrations of independent experiments should differ:[32] In a rectangular mole-fraction diagram, the two concentrations of one experiment are connected by a line and all lines of independent experiments intersect in one point and should be perpendicular to each other.

As 1D Raman spectra are convertible to concentration profiles of each species, multicomponent diffusion coefficients can be obtained from a single experiment.[73]

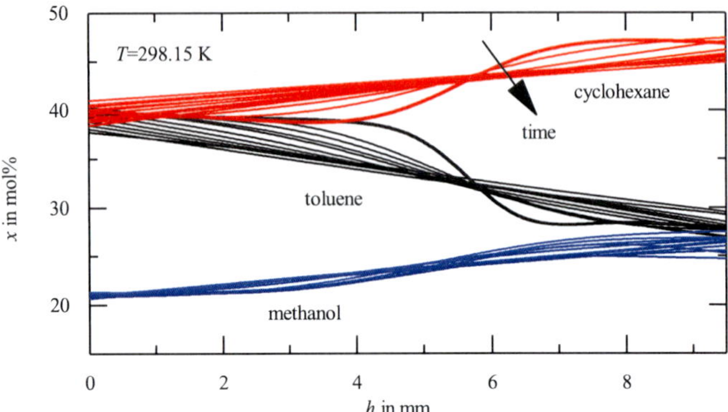

Figure 2.6: Typical profiles of the mole fractions x_i of cyclohexane + toluene + methanol along the cell height h over time during a ternary 1D Raman diffusion experiment. $h = 0$ corresponds to the bottom of the measurement cell and $h = 10\,\text{mm}$ to the upper end of the observed region.

Two independent experiments provide a ternary diffusion coefficient matrix with high accuracy and precision.[32]

Starting from this 1D Raman experiment, several variations and extensions have been published. Diffusion measurements using Raman spectroscopy have been adapted to higher temperatures (350 K) by this author and coworkers.[70] Diffusion measurements at increased pressures ($p \leq 40\,\text{MPa}$) were performed for one-dimensional diffusion of methane in water using a high-pressure optical cell.[83]

The diffusion coefficient of water in a carbohydrate glass was determined from spectra collected subsequently at various positions.[114] The temporally and spatially resolved loading of three solvents in film drying was used to estimate pseudo-binary diffusion coefficients.[115] Diffusion coefficients were also obtained in a similar procedure using other spectroscopic analysis, *e.g.*, UV-Vis in position scanning spectrophotometry.[116,117]

In general, diffusion measurements using (1D) Raman spectroscopy are suitable for organic as well as aqueous solutions with Raman active species. Raman spectroscopy

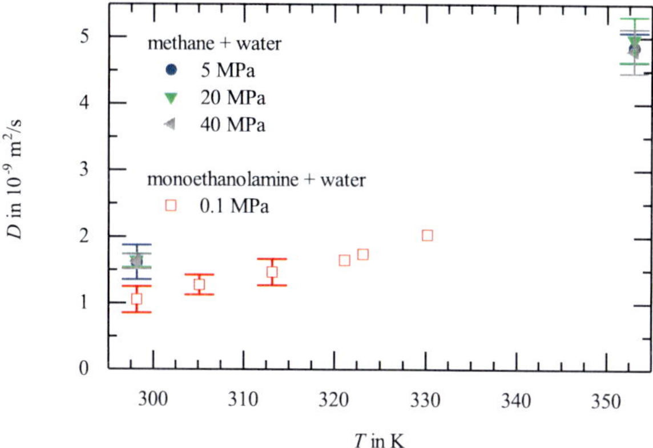

Figure 2.7: Diffusion coefficient of monoethanolamine + water as a function of temperature measured by our group (□).[70] Diffusion coefficients of methane + water (•, ▼, ◄) as a function of temperature and pressure measured by Guo *et al.*[83]

is interesting if the components of a system have distinctive peaks in their spectra and an adequate signal to noise ratio. To date, results of binary and ternary diffusion measurements using Raman spectroscopy have been published[32,70,73].

One disadvantage of the described diffusion measurements using 1D Raman spectroscopy is the batch character of the process: small initial disturbances have a large influence on the experiment. Also, diffusion starts with the initial contact of the samples during the experiment preparation. In the analytical solution, the time to the initial concentration profile is estimated together with the diffusion coefficient, but it also compensates for changing diffusion distances during layering of the solutions and, if necessary, the siphoning. Because there is a gas phase at the top of the diffusion cell, evaporation can induce a second concentration gradient. Due to the volume of the diffusion cell, heat of mixing can disturb the isothermal conditions. The experiment duration is slightly longer than for Taylor dispersion due to the comparatively large diffusion distance and the sample consumption is higher.

2.6 Microfluidics for Diffusion Measurements

Microfluidic diffusion measurements using Raman microspectroscopy benefit from the advantages of Raman spectroscopy, while the disadvantages specified above of previous 1D experiments are remedied.

The characteristic dimension of microfluidic devices ($< 1\,\mathrm{mm}$) allows for small sample consumption, short retention times, high throughput and steady state experiments.[33,118,119] Additionally, microfluidics profits from an intrinsically laminar flow regime and good heat exchange properties due to the large surface-to-volume ratio.[67,120] On the microscale, diffusion is the dominant mixing process because the diffusion distances are short.[18] Therefore, microfluidics is well suited for diffusion measurements.[10]

Microfluidics started in the early 1980s and is a rapidly growing field of science since the late 1990s.[6] Many macroscopic processes were transferred to microfluidics in the last decades, including unit operations such as distillation or extraction and devices such as pumps and valves.[121–124] Microfluidics is also used for particle sorting, droplet generation and manipulation, medical labs in the form of micro total analysis systems (μTAS), sensors based on micro-electro-mechanical systems (MEMS) and mixing processes as well as gradient generation based on diffusion.[7,37,125–128] The development of publications on microfluidics, and microfluidics together with diffusion is shown in Fig. 2.8. Approximately $10\,\%$ of the publications on microfluidics deal with diffusion.

In the following, microfluidic devices for diffusion measurements (section 2.6.1), characteristics of microfluidic flows (section 2.6.3) and the capabilities of microfluidic diffusion measurements (section 2.6.2) are highlighted.

2.6.1 Microfluidic Devices for Diffusion Measurements

Microfluidic chips and devices are multifunctional with customizable geometry and material which can adapt for diffusion measurements.[121] Glass is used prevalently as material for diffusion measurements because the advantages compensate for the disadvantages. While glass is reusable, chemically stable and optically accessible, it is also breakable and expensive.[6]

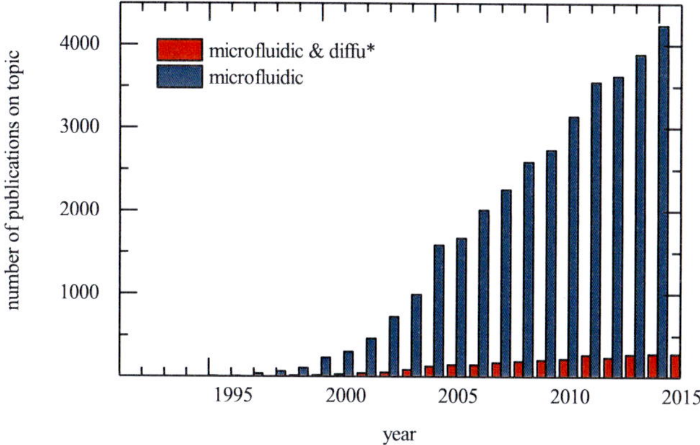

Figure 2.8: Development of the publications on the topics of microfluidics and diffusion since 1991. Data acquired from Web of Knowledge based on the search terms "microfluidics" and "microfluidics & diffusion".[129]

One common geometry for diffusion measurements is the H-cell, T-sensor, H-filter, or Y-Y-cell. The H-cell is based on a microchannel with two inlets and two outlets in form of a "Y", as shown in Fig. 2.9. The exact geometry of the H-cell strongly influences experiments, *i.e.*, flow characteristics: The inlet angle β of the H-cell should be small to reduce pressure drop, mixing and stagnant zones.[130] To reduce these effects, angles smaller than $90°$ are sufficient, while further reductions of the inlet angle show no significant influence.[8,131] Also, orifices should be avoided for diffusion measurements as they increase mixing and pressure loss.[131]

The volume flow and the ratio between the inlet flows influence the compositions at the outlet of the microchannel because mixing due to diffusion is dominant in microfluidics. Additionally, several channel inlets are possible to create specific concentration profiles. One example is sandwiching, which is the parallel flow of one core liquid between two flows of a second liquid.[14] If the microfluidic flow shall be sorted into n fractions, *i.e.*, sorted by concentration, n outlets are used. But precise mass splitting at the outlets is difficult to accomplish and requires knife-edges between the outlets.[132] To avoid splitting difficulties, *in situ* measurements are preferred.

The geometric ratio of the channel H/W with height H and width W influences the gravimetric stability and the simplifications in the mass transport model (section 2.6.3 and 2.6.2).[8]

To generate a volume flow, pressure-driven flow is widely used.[133] The implementation is comparatively easy and flexible.[134] Also, pressure-driven flow is insensitive to the ionic strength of liquids, the pH or surface contamination. A disadvantage of pressure-driven flow is the produced parabolic velocity profile in y- and z-direction, *i.e.*, a residence time distribution.[134]

To measure concentrations and determine subsequently diffusion coefficients, microfluidics can be combined with various analytics, *e.g.*, fluorescence spectroscopy, UV/Vis (absorption), or Raman spectroscopy.[9,76,121,135,136] The choice of analytics depends on the substances, their number and properties as well as the microfluidic chip material. Fluorescence spectroscopy or refractive index require additional effort for n-component systems, UV/Vis becomes complicated for organic solvents, glass as substrate is in-transparent for IR spectroscopy, while Raman spectroscopy is very applicable for microfluidics and n-component solvent systems.[6,27,86]

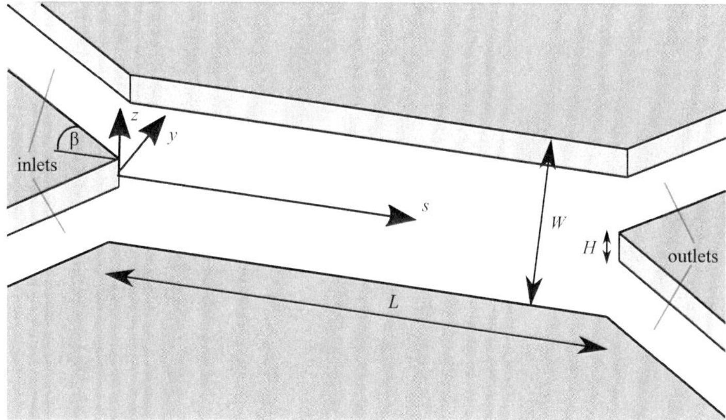

Figure 2.9: 3D sketch of the H-cell with inlet angle β, coordinate system s, y, z and width W, height H and length L. s denotes the axial direction, y the coordinate in width direction and z the coordinate in height direction. 2β is the angle between the two inlets of the H-cell.

2.6.2 Microfluidic Diffusion Measurements

As discussed above, microfluidics allows for short diffusion times due to short diffusion length and thus are advantageous for the measurement of diffusion.[5,31,36] In the 2000s, Kamholz *et al.* performed fundamental work on diffusion in microfluidics and microfluidic diffusion measurements.[10,134,137–139] Microfluidic diffusion measurements are based on the laminar co-flow of two miscible liquid phases in a microfluidic device, *e.g.*, an H-cell.[137] In these experiments, the liquid phases are contacted at the position where the channel inlets join ($s = 0$, $y = 0$) and diffusion occurs perpendicular to the convective, axial flow, see Fig. 2.10.

The H-cell is operated in steady state so that concentration measurements can be temporally integrated at the positions of interest.[134] The diffusion coefficient is estimated from comparison of experimentally determined concentration profiles with fitted concentration profiles from convection-diffusion models of the diffusion experiment, see 2.6.3.

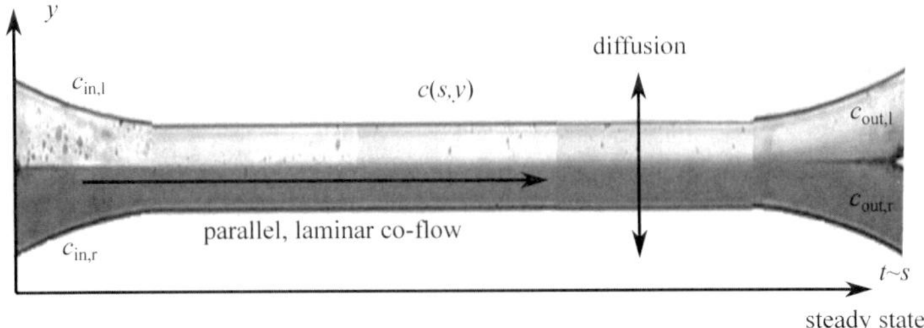

Figure 2.10: Directions of mass transport in a microfluidic diffusion experiment. The change of concentration is indicated by change of color.

The methods for concentration measurements exist *in situ*, inline at the channel outlet and offline behind the channel outlet. *In situ* concentration measurements allow spatially resolved measurements and rely on spectroscopic or optical analytics, *e.g.*, fluorescence, refractive index or Raman spectroscopy.[7,31,37,76,134,140] Fluorescence only provides information on the fluorescent component, the refractive index gives only one information for a mixture independent of the number of components n_c, while Raman spectroscopy offers information on all components as discussed in Sec. 2.5.1.

Inline concentration measurements at the outlet omit spatially resolved information due to the spatially integrated concentration measurements at all or chosen outlets. Inline concentration measurements use conductivity or UV/Vis measurements.[36,38]

The offline measurement of concentrations shows the same disadvantage as inline concentration measurements, *i.e.*, no spatial resolution. However, due to the free choice of analytics, *e.g.*, UV/Vis or refractive index, this method is feasible for all types of solutions.[132,141]

2.6.3 Modeling of Microfluidic Flows for Diffusion Experiments

The modeling of microfluidics and microfluidic diffusion measurements is based on the Navier-Stokes equations

$$\rho\left(\frac{\partial \boldsymbol{v}}{\partial t} + \boldsymbol{v}\nabla\boldsymbol{v}\right) = -\nabla p + \eta\nabla^2\boldsymbol{v} + \boldsymbol{f} \tag{2.23}$$

as conservation of momentum, with the body forces $\boldsymbol{f}$, fluid velocity field $\boldsymbol{v}$ and the shear viscosity η.[142] The Navier-Stokes equations often simplifies to the Stokes equation in microfluidics, as the inertial forces are small compared to the viscous forces and nonlinear terms are negligible[142]

$$\rho\frac{\partial \boldsymbol{v}}{\partial t} = -\nabla p + \eta\nabla^2\boldsymbol{v} + \boldsymbol{f} \ . \tag{2.24}$$

In the Stokes equation, the partial derivative of the mass flow is balanced by the gradient of the pressure, and friction and body forces.

In the modeling of diffusion, the conservation of mass is considered in the equation of continuity. The equation of continuity for component i in an incompressible fluid with n_c components is

$$\frac{\partial c_i}{\partial t} + \left(v_s\frac{\partial c_i}{\partial s} + v_y\frac{\partial c_i}{\partial y} + v_z\frac{\partial c_i}{\partial z}\right) = \sum_{j=1}^{n_c-1} D^V_{i,j}\left(\frac{\partial^2 c_j}{\partial s^2} + \frac{\partial^2 c_j}{\partial y^2} + \frac{\partial^2 c_j}{\partial z^2}\right) + R_i \tag{2.25}$$

$$\text{for } i = 1, ..., n_c - 1 \ .$$

Here, v_s, v_y and v_z are the constant velocities in s, y and z direction, and R_i the reaction rate of component i.[51,139] Diffusive transport of i occurs due to gradients of c_j with the diffusion coefficient matrix $\boldsymbol{D}^V$. The presented equation of continuity assumes constant elements of the diffusion coefficient matrix $D^V_{i,j}$.

The mass transport modeling of microfluidic diffusion experiments generally assumes laminar flow, isothermal conditions, steady state, and a negligible entrance flow region. The limitations of these assumptions are briefly discussed in the following paragraphs.

In most applications, microfluidic flows are laminar, *i.e.*, the streamlines are parallel to each other and to the surface ($v_y = v_z = 0$) and no exchange between parallel layers occurs.[134,143–145] Laminar flows are characterized by a small Reynolds number

$$Re = \frac{\rho v d}{\eta} \ll 100 \ , \tag{2.26}$$

which describes the ratio of inertial forces to viscous forces. Here, d denotes the characteristic diameter. For a large Reynolds number ($Re > 100$), the Navier-Stokes equations is necessary instead of the Stokes equation. The critical Reynolds number Re_c is the lower limit for turbulent flow and depends on the microfluidic geometry, because the influence of roughness is negligible. For a ratio $L/d_H \geq 70$ with length L and the hydraulic diameter d_H, the critical Reynolds number remains $Re_c = 2300$.[124] Chen *et al.*[146] have observed three different regimes of the critical Reynolds number as shown in Fig. 2.11. If turbulence is desirable in microfluidic flows at low Re, it has to be enforced by electrokinetics or Dean flows resulting from the geometry.[144,147]

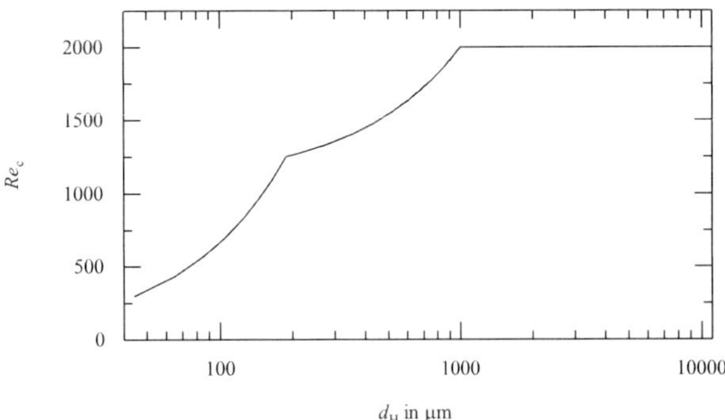

Figure 2.11: Critical Reynolds number Re_c as function of the hydraulic diameter of the microfluidic device as described by Chen *et al.*[146]

Behind the entrance flow region, the flow is hydrodynamically fully developed. For a flat channel at the macroscale, the length of the entrance flow region L_f is[148]

$$\frac{L_\mathrm{f}}{H} = \frac{0.45}{1 + 0.41Re/0.45} + 0.41Re \ , \tag{2.27}$$

with the channel height H. An empirical correlation for the entrance flow region in microfluidic devices is[138]

$$L_\mathrm{f} = d_\mathrm{h}(0.379e^{-0.148Re} + 0.055Re + 0.260) \ . \tag{2.28}$$

In microfluidics, the length of the entrance flow region L_f is usually shorter than the dimension of the channel width W and negligible for long channels with $L/W \gg 1$.[76,149,150]

The thermal entry length L_e is the point after which a fluids initial temperature cannot be retraced. It is determined according to

$$L_\mathrm{e} \sim \frac{1}{16}\frac{vW^2}{\kappa} \tag{2.29}$$

using the thermal diffusivity κ.[67] In practice, the thermal entry length is $L_\mathrm{e} \leq 10W$.

Steady state was observed to be easily accessible in microfluidic flow-through devices.[136,150,151] The time to steady state can be approximated with $\tau_\mathrm{e} \sim 10\,\mathrm{ms}$ for a $100\,\mu\mathrm{m}$ microchannel, *i.e.*, the time to diffuse short distances as the channel width W.[142] Another approximation for the time to steady state is[142]

$$\tau_\mathrm{e} \sim \frac{\rho W^2}{\eta} \ . \tag{2.30}$$

Due to the pressure-driven flow in microfluidic diffusion experiments, parabolic velocity profiles similar to macroscopic laminar flow in a duct are observed.[152] The parabolic velocity profile approaches plug flow in y-direction for $W/H > 4$ while it remains parabolic in z-direction in case of constant viscosities and fully developed flow in rectangular duct.[138,149] For a geometric ratio $W/H > 20$ this results in a constant velocity $v(y)$ for 90 % of the channel width.[134] Phenomena like slip flow are limited to non-Newtonian fluids in microfluidics.[153]

Due to the small dimension in microfluidics, the pressure drop is large for typical settings. For a rectangular micro channel with a geometric ratio $W/H > 10$, the pressure drop is[67,154]

$$\Delta p = \frac{12L\eta}{H^3W}\dot{V} \,. \tag{2.31}$$

This results in a pressure drop per length of

$$\frac{\Delta p}{L} = \frac{12\eta}{H^3W}\dot{V} = 7\,\mathrm{bar\,m^{-1}} \tag{2.32}$$

for water at 298.15 K in a microfluidic device with $H = 40\,\mu\mathrm{m}$ and width $W = 400\,\mu\mathrm{m}$ at a volume flow of $\dot{V} = 0.1\,\mathrm{mL\,min^{-1}}$.

For spiral-coiled or meandering microchannels, the centrifugal effects due to the curvature have to be considered. The centrifugal effects induce secondary flow-fields, such as annular flows and phase switch, and increase mixing.[142,155] Therefore, significant centrifugal effects should be avoided for microfluidic diffusion experiments. The centrifugal effects due to curvature are characterized by the Dean number

$$De = \left(\frac{d_\mathrm{h}}{d_\mathrm{curvature}}\right)^{0.5} Re \tag{2.33}$$

describing the ratio of inertial and centrifugal forces to viscous forces.[147,156] d_h denotes the hydraulic diameter and $d_\mathrm{curvature}$ the diameter of the spiral or the meander. For geometric ratios $W/d_\mathrm{curvature} < 4$ with the channel width W, secondary flows almost disappear.[121] Song and Cabooter[25] observed no influence from secondary flows for $De^2Sc \ll 100$, where the Schmidt number

$$Sc = \frac{\eta}{\rho D} \tag{2.34}$$

describes the ratio of viscous diffusion rate to mass diffusion rate.

The dynamic viscosity η is important for diffusion experiments, as microfluidic flows are controlled by viscous forces suppressing convective mixing.[33,152] However, if fluids of different viscosities are contacted, the viscosities influence the velocities of the fluids and the occupied cross-section.[157] For example, for two Newtonian liquids of different viscosities but equal volume flow, the more viscous liquid has a lower velocity and fills the larger cross-section according to

$$\frac{\eta_\mathrm{l}}{\eta_\mathrm{r}} = \frac{1/2 + Y}{1/2 - Y} , \tag{2.35}$$

where Y is the dimensionless position of the interface, and the indices l and r denote left and right side of the channel, respectively.[158] In microfluidic diffusion experiments, constant viscosities are often assumed, *i.e.*, negligible viscosity changes can be assumed for the concentration difference of interest.[133] However, the validity of this assumption needs to be checked and non-Newtonian fluids generally behave differently.[121,159] Details and consequences for microfluidic diffusion measurements are discussed in the following paragraphs.

In diffusion experiments, two liquid phases co-flow, either on top of each other or next to each other. Thus, the influence of the density is of interest. In a first assumption, the interface is stable independent of the spatial orientation.[13] However, a retention-time dependent reorientation of the interface for convection dominated systems was observed by Yoon *et al.*[160] For a quadratic channel cross-section, the inversion width of the interface $\zeta(s)$ due to reorientation is empirically described as

$$\zeta(s)/W = 0.075 H^{0.705} (g \Delta\rho)^{0.595} s^{0.51} \rho^{0.025} v^{-0.57} \eta^{-0.62} \tag{2.36}$$

at axial position s, with the gravity acceleration g. Lin *et al.* suggest reducing the height of the diffusion device as the absolute height dominates the interface reorientation for a given chemical system.[161] The reorientation of the interface should be as small as possible for diffusion experiments. To reduce the complexity of the mass transport models, the models often assume negligible excess volumes ($V^\mathrm{E} = 0$) or a negligible change of the excess volume in the observed concentration range ($\Delta V^\mathrm{E} = 0$) because these excess volumes change the velocities.[76,157]

Due to the laminar flow and $v_y^V = v_z^V = 0$ in the mass conservation (2.25), diffusion is the only mixing process in microfluidics in y- and z-direction.[4,33,152] In s-direction, the mass conservation (2.25) specifies that the diffusive mixing decreases by increasing velocity.[120] The ratio of convective to diffusive transport is described by the Péclet number

$$Pe = \frac{Wv}{D} \, , \tag{2.37}$$

i.e., the Péclet number Pe indicates after how many channel widths downstream complete mixing is achieved.[142]

The diffusion time reduces quadratically with diffusion distance as indicated by the conservation of mass (2.25).[8,18] Therefore, diffusive broadening in microfluidics is often described phenomenological by a power law[120,134,142]

$$\delta(s) \sim \left(\frac{Ds}{v}\right)^{e} \, . \tag{2.38}$$

Here, $\delta(s)$ denotes the width of the diffusion zone at position s. While the exponent $e = 1/2$ is expected, this exponent is observed only at the half-depth plane ($z = 0$) and integrated over the channel height ($\int_{-H/2}^{H/2} \delta(s)dz$). Due to the velocity distribution, a phenomenological power $e = 1/3$ is observed at the channel wall and the exponent changes from the wall to the half-depth plane.[120,134,142] This phenomenon is observed in the butterfly effect: An increased residence time near the wall due to the parabolic velocity profile results in a more wide-spread diffusion zone, leading to artifacts in diffusion measurements. The effect disappears for $W/H \gg 1$, $s/(HPe) \sim 0.1$ and $s \gg vH^2/D$.[76,150,162] In all cases, the equation of continuity (2.25) is satisfied.

Axial diffusion, also described in the equation of continuity (2.25), complicates the microfluidic measurement of diffusion. Axial diffusion is important especially for small Péclet numbers and is increased by Taylor-dispersion.[67,159,163] To neglect axial diffusion, large Péclet numbers $Pe = vW/D \gg 1/(L/W)$ are recommended, *e.g.*, 150.[120] For example, Beard observed for $L/W = 3$ only small deviations between a model considering and a model neglecting axial and Taylor dispersion for $Pe = 1500$ and very good agreement for $Pe = 15000$.[164] Also, axial diffusion is negligible locally for $vs/D \gg 1$.[162]

Diffusion in z-direction becomes important for $H/W > 0.45$.[131] For $Pe \gg s/H \gg 1$, the diffusive broadening $\delta = \delta(s, z)$ also depends on z due to the parabolic velocity profile.[120] For $s/H \gg Pe \gg 1$, the diffusive broadening $\delta = \delta(s)$ depends only on the axial distance s because the diffusive transport in z is fast enough to neglect

$\partial c/\partial z$.[120] The influence of the channel wall on diffusion is only observed close to the wall (1 µm).

2.6.4 Convection-Diffusion Models for Microfluidic Diffusion Experiments

Convection-diffusion models are required to determine diffusion coefficients from measured concentrations. These convection-diffusion models are generally based on the (Navier-)Stokes equations ((2.23), (2.24)) and the equation of continuity (2.25) with the considerations discussed in the preceding section 2.6.3. Available models range from 3D diffusion-convection approaches solved numerically *via* finite element method to analytical solutions for 1D free diffusion.[7,18,38,76,165–167,167,168] In between, several levels of simplification exist, *e.g.*, 3D convection combined with 2D diffusion or a 1D velocity gradient combined with 1D or 2D diffusion.[120,127,133,134,143,149,150,159] While numerical calculations introduce numerical diffusion due to coarse meshes, for many of the above models, no analytical solution is known or possible.

Here, the 2D convection-diffusion model presented by Häusler *et al.*[36] is highlighted. This steady state model

$$\bar{v}(y)\frac{\partial c}{\partial s} = D^V \frac{\partial^2 c}{\partial y^2} \tag{2.39}$$

is suited for non-reactive binary systems in restricted y-diffusion geometries. The velocities in y- and z-direction are zero ($v_y^V = v_z^V = 0$), the velocity in s-direction is $v_s^V = v$. For $-W/2 < y < W/2$ the velocity is integrated over the channel height resulting in the velocity in s-direction $\bar{v}(y)$. Only diffusion in y-direction is considered as axial diffusion and diffusion in z-direction are assumed to be negligible.

The initial conditions describe a jump of the concentration at the channel inlet

$$\begin{aligned} c &= 0 \;\forall\;\; y < 0 \;, \\ c &= 1 \;\forall\;\; y \geq 0 \;. \end{aligned} \tag{2.40}$$

Here, the concentrations are normalized, *i.e.*, $c_{\min} \equiv 0$ and $c_{\max} \equiv 1$. The no-flux boundary conditions

$$\frac{\partial c}{\partial y}\bigg|_{y=y_{\min};y=y_{\max}} = 0 \tag{2.41}$$

complete the convection-diffusion model.

For microfluidic diffusion measurements, 1D models together with free diffusion are dominantly used. For free diffusion models, the Fourier number is $Fo \sim 0$: The Fourier number

$$Fo = \frac{\tau \cdot D^*}{(W/2)^2} , \tag{2.42}$$

describes the ratio of the contact time τ to the characteristic diffusion time $(W/2)^2/D^*$ with the geometric mean of the diffusion coefficient D^*. However, diffusion measurements in restricted diffusion geometries provide more information per experiment. For rectangular microchannels, the optimal contact time regarding information on the diffusion coefficient depends on the geometry and is modeled as[36]

$$Fo_{\mathrm{opt}} = 0.299 \left(\frac{W}{H}\right)^{-0.0983} \quad \text{for} \quad 0.05 < H/W < 1 . \tag{2.43}$$

Thus, for many geometries the Fourier number is in the range $Fo = 0.3 ... 0.4$.

2.6.5 Capabilities of Microfluidic Diffusion Measurements

Microfluidic diffusion measurements have already been used for systems containing bio-molecules,[10,137,138,169,170] dye molecules,[8,18,35,38,127,137,171] sugars,[37,39,132,149] electrolytes[36,39,166] and alcohols.[18,35,149,166,168,169] Though microfluidic diffusion measurements are flexible in their application, most systems were aqueous.[8,18,35–39,127,132,166,168,169,171] Organic solvent systems are the exception and these exceptions have typically already frequently been published such as cyclohexane + toluene.[70,76,78,172]

In addition to the highlighted microfluidic diffusion measurement methods using H-cells, a variety of microfluidic approaches exists, *e.g.*, using modifications of the geometry (T-cell, "'+"'-inlet),[14,41,133,166,168] relating viscosity and diffusion coefficient,[138,173] microfluidic Taylor dispersion and plug-flow experiments,[35,37,169] or batch experiments between reservoirs.[171]

On the one hand, microfluidic diffusion measurements suffer from general disadvantages of microfluidics, *e.g.*, fouling, blocking, and breakable glass. On the other hand, microfluidic diffusion measurements benefit from short contact times (<1 min), short experiment time (5 min to 45 min), and small sample volumes (1 µL to 10 µL).[10,10,36,39,76,137,138,174] Though isothermal operation is difficult, isothermal diffusion measurements at various temperatures are possible with suitable peripheral devices.[76,175] Measurement accuracy and precision cover a wide range, from low precision or accuracy for free diffusion experiments to high precision and accuracy ($> 99\,\%$) for diffusion in a restricted geometry.[36,38,139,166,170]

Overall, microfluidics shows a large potential for the measurement of multicomponent diffusion coefficients.[5,31,36] Still, microfluidic multicomponent diffusion measurements have not been accomplished yet.

2.7 Contribution of this Thesis

Measurements of multicomponent diffusion coefficients are indispensable. Nevertheless, the literature review revealed drawbacks of classical diffusion measurements but also current limitations for diffusion measurements based on Raman and microfluidics. To date, Taylor dispersion is the method of choice for diffusion measurements, because multicomponent diffusion measurements are feasible, and the experiment time is short. But Taylor dispersion or other classical diffusion measurements require at least $n_c - 1$ independent experiments for a multicomponent diffusion coefficient. In practice, many more experiments are performed, *e.g.*, 5 ternary experiments plus repetitions, or 16 quaternary experiments.[88,176] Therefore, the aim of this thesis is to develop and validate a new method for diffusion measurements suited for multicomponent systems to reduce the measurement effort and fill the diffusion data gap. For this purpose, microfluidics and Raman microspectroscopy are combined to measure binary and multicomponent diffusion coefficients in very few experiments and with short experiment time. The contribution of this thesis is divided into the development of the diffusion measurement method, the validation of the method and the measurement of new diffusion data.

Raman microspectroscopy and microfluidics for the measurement of multicomponent diffusion coefficients.
In this thesis, a measurement method for multicomponent diffusion coefficients is developed based on Raman microspectroscopy and microfluidics (Chapter 3). For this purpose, design guidelines for microchannels and the experimental procedure are derived from the literature review and implemented in an experimental setup and an experimental procedure. The experimental procedure is also founded on results from model-based experimental design. A convection-diffusion model for multicomponent diffusion in microfluidics is formulated. This convection-diffusion model and the modeling of diffusive mass transport are the base for the developed data analysis procedure for diffusion coefficients.

Validation of multicomponent diffusion measurements using Raman microspectroscopy and microfluidics.
Binary and ternary systems for which diffusion data is available from literature were chosen to validate the developed method (section 3.1). These systems are the binary mixtures cyclohexane + toluene, acetone + water, and acetone + toluene, and the ternary mixtures 1-propanol + 1-chlorobutane + heptane, and cyclohexane + toluene + methanol. For these systems, the diffusion coefficient is obtained from a least-squares procedure fitting the measured concentrations to the convection-diffusion model. The agreement between measured and fitted concentrations defines whether the convection-diffusion model is suited to describe the experiment and the diffusion coefficient can be plausible (Chapter 4). The determined diffusion coefficient is discussed in comparison to the diffusion coefficient (matrix) from literature.

Measurement of new multicomponent diffusion data using Raman microspectroscopy and microfluidics.
Because diffusion measurements are time-consuming and difficult, diffusion data is missing for many multicomponent systems but also for binary systems. In this thesis, new data was measured for the binary systems cyclohexane + methanol, the ternary system water + acetone + toluene, and the quaternary system cyclohexane + toluene + acetone + methanol. Though water + acetone + toluene is a recommended test system for liquid-liquid extraction, multicomponent diffusion data is missing.[177] For quaternary systems of organic solvents, there is generally a lack of diffusion data (Fig. 2.2). This new diffusion data is compared to predictions from engineering models. The results are discussed in Chapter 4.

Chapter 3

Towards Diffusion Measurements Using Raman Microspectroscopy and Microfluidics

Major parts of this chapter are adapted from Ref.[17] with permission from The Royal Society of Chemistry:

> C. Peters, L. Wolff, S. Haase, J. Thien, T. Brands, H.-J. Koß, A. Bardow (2017). Multicomponent diffusion coefficients from microfluidics using Raman microspectroscopy. *Lab Chip*, vol. 17, pp. 2768.

Contribution report: Writing the draft, principal author, planning the experimental setup, choosing the chemical systems, planning the experiments, supporting experiments conducted by S. Haase and R. Becka, data evaluation.
Parts of this chapter are adapted with permission from

> C. Peters, J. Thien, L. Wolff, H.-J. Koß, A. Bardow (2019). Quaternary Diffusion Coefficients in Liquids from Microfluidics and Raman Microspectroscopy: Cyclohexane + Toluene + Acetone + Methanol. *J. Chem. Eng. Data*, `https://doi.org/10.1021/acs.jced.9b00632`.[42] Copyright (2019) American Chemical Society.

Contribution report: Writing the draft, principal author, planning the experimental setup, choosing the chemical systems, planning the experiments, supporting experiments conducted by J. Thien and R. Becka, data evaluation.

Model based experimental design and parameter estimation procedure were developed in cooperation with Ludger Wolff.

In this chapter, the investigated chemical systems are presented together with the experimental setup for microfluidic Raman diffusion measurements, the applied convection-diffusion model, the resulting experimental procedure and the data analysis as well as the parameter estimation. Additionally, the procedure is outlined for the prediction of diffusion coefficients based on engineering models.

3.1 Chemical Systems

In this work, mutual diffusion coefficients were measured in binary, ternary and quaternary systems. The binary systems are the non-aqueous systems cyclohexane + toluene, acetone + toluene, and cyclohexane + methanol as well as the aqueous system acetone + water. The ternary systems are cyclohexane + toluene + methanol, 1-propanol + 1-chlorobutane + heptane, and the aqueous system water + acetone + toluene. The quaternary system is cyclohexane + toluene + acetone + methanol. Criteria for the choice of systems are whether the substances are Raman active and the Raman spectra are distinguishable, toxicity, and, if possible, availability of literature data on diffusion for comparison. Pure component data is shown in Table 3.1. According to Table 3.1, all components have distinctive spectral bands within one system.

All binary systems show deviations from ideal mixing behavior regarding excess volumes, enthalpy of mixing, or azeotropic type (Table 3.2). Table 3.2 also shows available literature data on diffusion coefficients in the binary systems.

Cyclohexane + toluene is particularly interesting, as it has been repeatedly measured by various groups.[76,78,172,204] The enthalpy of mixing shows large deviations from ideal mixing behavior in comparison to the other investigated systems (Table 3.2). While both substances have similar molar volumes, viscosities differ significantly though the relative difference is small compared to the other systems.

Cyclohexane + methanol is studied as the second subsystem of cyclohexane +

Table 3.1: Molar volume V_i^0, dynamic viscosity η_i, thermal conductivity κ_i, and spectral bands of used substances

	V_i^0 in $10^{-6}\,m^3\,mol^{-1}$	η_i in mPa s	κ_i in $W\,m^{-1}\,K^{-1}$	spectral bands in cm^{-1}
acetone[178–180]	74.05	0.31	0.17	787, 1530, 1708, 2921
1-chlorobutane[181–183]	105.2	0.42	0.12	652, 732, 814, 875, 1052, 1109, 1452
cyclohexane[184–186]	108.8	0.89	0.12	803, 1029, 1267, 1446, 2853, 2938
heptane[187–189]	147.4	0.39	0.12	855, 902, 1046, 1140, 1303, 1461
methanol[190–192]	40.74	0.55	0.20	1037, 1453, 2935, 2945
1-propanol[193–195]	75.17	2.0	0.15	861, 889, 970, 1050, 1106, 1275, 1290, 1460
toluene[196–198]	106.9	0.56	0.13	522, 787, 1005, 1032, 1211, 2930, 3066
water[178,199,200]	18.07	0.89	0.61	3266, 3401, 3616

Table 3.2: Excess volumes V^{E}, excess enthalpies h^{E}, relative difference in molar volume and dynamic viscosity, azeotropic type of investigated binary systems, and literature on diffusion coefficients D

	$V^{\mathrm{E}}_{\mathrm{min}}$ in $\mathrm{cm^3\,mol^{-1}}$	$V^{\mathrm{E}}_{\mathrm{max}}$	$h^{\mathrm{E}}_{\mathrm{min}}$ in $\mathrm{J\,mol^{-1}}$	$h^{\mathrm{E}}_{\mathrm{max}}$	$\Delta V/V_{\mathrm{min}}$ in %	$\Delta\eta/\eta_{\mathrm{min}}$ in %	azeotropic type	references on D
cyclohexane + toluene[201–203]		0.59		628	2	59	zeotropic	[76,78,172,204]
cyclohexane + methanol[205–207]		0.50		538	167	62	pressure max.	a, [47,208]
acetone + toluene[209–211]	−0.16			244	44	81	zeotropic	[212]
acetone + water[213–215]	−1.5		−684	320	310	97	zeotropic	[216–218]

[a] only cyclohexane in methanol at infinite dilution

toluene + methanol and the quaternary system cyclohexane + toluene + acetone + methanol. For cyclohexane + methanol, diffusion coefficients were only available for cyclohexane in methanol at infinite dilution.[47,208] Diffusion coefficients for methanol in cyclohexane at infinite dilution have recently been published.[219] In this work, additional diffusion coefficients of cyclohexane in methanol were measured, which allow comparison to predictive engineering models. Predictions are presumably challenging because the system shows strong deviations from ideal mixing behavior, *i.e.*, a miscibility gap and an azeotrope (Table 3.2).[220]

Acetone + toluene and acetone + water are fully miscible subsystems of the ternary test system for liquid-liquid extraction acetone + toluene + water. While acetone + toluene shows small deviations from ideal mixing behavior, acetone + water shows comparably large excess volumes, a change of sign in the heat of mixing, and large relative differences between molar volumes and between dynamic viscosities (Table 3.2).

In the ternary system 1-propanol + 1-chlorobutane + heptane, all pure component molar volumes differ significantly, and the viscosity of 1-propanol is significantly larger than the viscosities of 1-chlorobutane and heptane (413 % and 8 %, Table 3.1). The system shows non-ideal mixing behavior but is still completely miscible.[2] Though the Raman spectra have several overlapping bands in the fingerprint region, the spectra are still distinguishable (see Fig. 4.16). Literature data on diffusion is available to compare results from this work[2,32]

The ternary system cyclohexane + toluene + methanol shows non-ideal mixing behavior, and there is a miscibility gap in the low toluene region.[221] Even in the ternary mixture, the Raman spectra are sufficiently distinguishable in the fingerprint-region. Also, diffusion coefficients have repeatedly been measured for cyclohexane + toluene + methanol[1,222,223] and predicted using MD.[224]

The ternary system water + acetone + toluene is a test system for liquid-liquid extraction.[177] Nevertheless, only pseudo-binary diffusion coefficients of the solute have been measured yet.[225] The system shows a large miscibility gap between toluene and water, ranging to acetone contents of 25 wt%. The Raman spectra are well distinguishable in the methyl group- and OH-region for large water contents.

As quaternary system, cyclohexane + toluene + acetone + methanol was chosen. All components are Raman active in this system (and distinguishable), in contrast to most of the published quaternary systems.[22,23] Additionally, the system is not carcinogenic, all substances are liquid at 25 °C, the system includes only widely-used chemicals, and for all but one binary subsystems diffusion coefficients are available from literature.[226]

All chemicals were used as received, see Table 3.3.

Table 3.3: Used chemicals with specification

	supplier	description	purity percent
acetone	Merck Millipore	Uvasol	99.9
1-chlorobutane	Merck	LiChrosolv	99.8
cyclohexane	VWR	Spectronorm	99.7
heptane	Bernd Kraft	p.a.	99
methanol	VWR	Spectronorm	99.9
1-propanol	Merck	LiChrosolv	99.8
toluene	VWR	Spectronorm	99.8
water	Merck	SupraSolv	99

3.2 Experimental Setup

The experimental setup is based on Raman microspectroscopy in an H-cell microchannel (Fig. 3.1). In the H-cell, two solutions of different compositions are contacted, and the compositions equilibrate through diffusion perpendicular to the flow direction.

The microfluidic chip (Micronit, Netherlands) consists of two bonded borosilicate glass plates. The thickness of the upper and lower glass is 1000 µm and 700 µm, respectively. The H-cell is wet-etched into the lower of the two glass plates. The H-cell is a custom-made channel in form of a spiral of length $L = 1$ m, width $W = 400$ µm, and height $H = 40$ µm (Fig. 3.1B), C)). Advantages of microfluidic chips made from glass are their reusability and durability in severe chemical environments and organic solvents.

The temperature of the microfluidic chip is controlled via a water-flushed aluminum plate on top of the microfluidic chip. Additionally, glycerol increases the heat

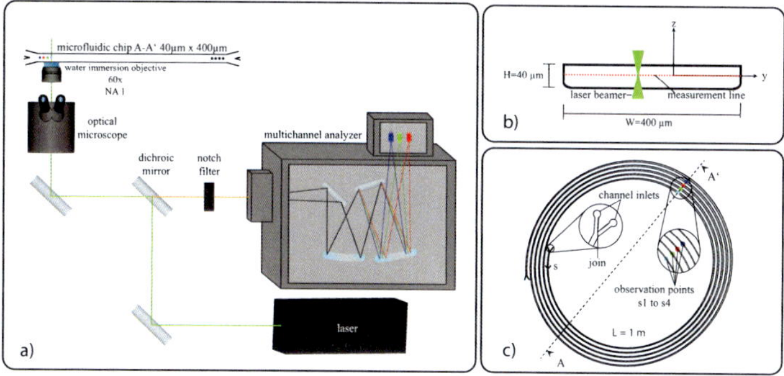

Figure 3.1: A) Optical setup consisting of laser, microfluidic chip, optical microscope, and multichannel analyzer. B) Channel cross-section with dimensions, coordinate system, laser beam, confocal volume, and measurement line (+). C) Top view of microfluidic chip with microchannel in form of a spiral H-cell with observation points s_1 (–), s_2 (–), s_3 (–), and s_4 (–). The enlarged circle shows the inlet of the H-cell.

exchange between the microfluidic chip and the water-flushed aluminum plate in comparison to air. In the aluminum plate, a Pt 100-sensor connected to a thermostat (CC-K15 Pilot One, Peter HUBER Kältemaschinenbau AG, Germany) measures the temperature. This temperature is the control parameter for the thermostat, which provides water as cooling liquid. Additionally, the temperature below the microfluidic chip is measured with a thermocouple. Thereby, a temperature precision of $\pm 0.5\,\mathrm{K}$ is achieved in the microfluidic chip.

The dosing system consists of a push-pull syringe pump (PHD ULTRA, Harvard Apparatus, US) with two 2.5 mL syringes (H-TLT, ILS Innovative Labor Systeme GmbH, Germany) and syringe filters with a pore diameter of 0.45 µm. For polar systems, polyamide filters and for non-polar systems PTFE filters are used (MULTOCLEAR-PA and -PTFE, CS-Chromatographie Service GmbH, Germany). Tubing from PTFE ($D_i = 0.3\,\mathrm{mm}$) and connecting parts from PEEK complete the system.

An inverse confocal Raman microscope (inVia, Renishaw, United Kingdom) is equipped with a frequency-doubled Nd:YAG-laser of 100 mW at 532 nm. The microscope is operated with a water immersion objective (NA = 1, WD = 2 mm, LUMPLFLN 60XW, Olympus, Japan). The attached multichannel analyzer consists of a spectrometer with a grating of 1800 grooves/mm and a cooled CCD-camera (1024x256 Pixel). The opening of the slit in front of the CCD chip is 65 µm, while the readout area of the CCD-Chip consists of three binned lines. This configuration results in a confocal volume of 2 µm x 2 µm x 10 µm (l x w x h) (Fig. 3.1B)). A smaller confocal volume of 1 µm x 1 µm x 5 µm according to Sec. 2.5 is possible for a smaller readout area of the CCD-Chip and a smaller opening of the slit. This smaller confocal volume would increase the exposure time multiple times. However, the above-mentioned configuration allows short experiment durations and adequate spatial resolution for diffusion measurements.

3.3 Convection-Diffusion Model

The model-based estimation procedure for the diffusion coefficients is based on a mass transport model. For this purpose, the convection-diffusion model for binary mixtures developed by Häusler *et al.*[36] (eqn (2.39)-(2.41)) is extended to multicomponent mixtures in this work.

A no-slip, laminar flow in a duct with a fully developed and gravimetrically stable flow field is assumed, see Sec. 2.6. These assumptions are based on the fact that the channel is flat ($W \gg H$, see Fig. 3.1B)) and that the magnitude of the Reynolds number Re is of order 1. Multicomponent diffusion equilibrates the concentrations perpendicular to the direction of flow.

The diffusion process is continuous and in steady-state. Therefore, the length coordinate s (Fig. 3.1C)) is proportional to an average contact time. Convection dominates the mass transport in length-direction s (Péclet number $Pe > 100$), while diffusion dominates the mass transport in the y-z cross-sectional plane. Furthermore, we average the concentrations over the z-direction because concentrations are smoothed out by diffusion much faster in z-direction than in y-direction due to the high aspect ratio ($W \gg H$).

The channel is fed with solutions composed of mole fractions $x_{\mathrm{l},i}$ and $x_{\mathrm{r},i}$ of component i for the left (l) and right (r) channel inlet, respectively, *i.e.*,

$$x_i(s=0,y) = \begin{cases} x_{\mathrm{l},i} & \text{for } y < y_0 \\ x_{\mathrm{r},i} & \text{for } y \geq y_0 \end{cases}, \qquad i = 1, ..., n_c - 1. \tag{3.1}$$

Here, y_0 denotes the position where the two channel inlets join. The difference in mole fractions $\Delta x_i = |x_{\mathrm{l},i} - x_{\mathrm{r},i}|$ between the left and right channel inlet is chosen as small as possible while allowing for distinguishable concentration measurements in this interval. Thereby, constant viscosities and negligible excess volume effects can be assumed within the mole fraction difference Δx_i. Over the measured mole fraction range Δx_i, the diffusion coefficient matrix $\boldsymbol{D}$ is also averaged. For the small Δx_i employed here, a constant diffusion coefficient matrix $\boldsymbol{D}$ is assumed as the change of the diffusion coefficient is small over a small mole fraction range. Wolff *et al.* have shown that for small changes of the diffusion coefficient the error is negligible.[227]

Using generalized Fick's law to describe the diffusive fluxes,[3] the multicomponent convection-diffusion model then reads:

$$v(y) \cdot \frac{\partial c_i(s,y)}{\partial s} = \sum_{j=1}^{n_c-1} D^V_{i,j} \frac{\partial^2 c_j(s,y)}{\partial y^2} \quad \text{for } i = 1, ..., n_c - 1, \tag{3.2}$$

$$x_i(s,y) = \frac{c_i(s,y)}{c_t(s,y)} \quad \text{for } i = 1, ..., n_c, \quad c_t(s,y) = \sum_{i=1}^{n_c} c_i(s,y) \quad \text{and}$$

$$\frac{1}{c_t(s,y)} = \sum_{i=1}^{n_c} x_i(s,y) \cdot V_i^0 .$$

Here, c_i denotes the molar concentration averaged over the z-coordinate, x_i the mole fraction, c_t the total molar concentration, and V_i^0 the molar volume of component i. The literature data on pure component molar volumes V_i^0 are listed in Table 3.1.

Instead of the pure component molar volumes V_i^0, partial molar volumes $\bar{V}_i$ are used for all binary systems, cyclohexane + toluene + methanol, and cyclohexane + toluene + acetone + methanol (Sec. D-G, I, K).

For 1-propanol + 1-chlorobutane + heptane, the difference between partial molar volumes and pure component molar volumes was negligible (0.4 %, Sec. H). For water + acetone + toluene, the measured data on densities in the mixture and the determined partial molar volumes was inconclusive, *i.e.*, the excess volume shows fluctuations and no clear concentration dependence (V^E/V ranges from $-4\,\%$ to $6\,\%$). Due to reliable data on V_i^0 (Table 3.1), the pure component molar volumes were used for 1-propanol + 1-chlorobutane + heptane and water + acetone + toluene. The elements $D_{i,j}^V$ of the $(n_c - 1) \times (n_c - 1)$ diffusion coefficient matrix $\boldsymbol{D}^V$ are determined in the volumetric reference frame.

The z-averaged velocity profile $v(y)$ is necessary, as even for $W > 20H$ the velocity is constant only over $90\,\%$ of the width, see Sec. 2.6.3. For a laminar, no-slip flow in a rectangular duct the velocity profile is given by[228]

$$v(y) = \left(\frac{m+1}{m}\right)\left[1 - \left(\frac{y}{(W/2)}\right)^m\right] \cdot \bar{v} \tag{3.3}$$
$$\text{with } m = 1.7 + 0.5\left(\frac{H}{W}\right)^{-1.4}.$$

Here, $\bar{v}$ denotes the average velocity in the channel given by $\bar{v} = \dot{V}/(H \cdot W)$ with the total volume flow in the channel $\dot{V}$.

In addition to the inlet conditions (3.1), the convection-diffusion equation (3.2) is complemented by the two no-flux boundary conditions at the channel walls:

$$\frac{\partial c_i(s, -W/2)}{\partial y} = \frac{\partial c_i(s, W/2)}{\partial y} = 0, \quad i = 1, ..., n_c - 1. \tag{3.4}$$

The progress of diffusional mixing can be described by the dimensionless Fourier number *Fo* (2.42). The Fourier number relates the contact time τ to the characteristic diffusion time $(W/2)^2/D^*$. Generally, D^* is a characteristic value of the diffusion coefficient. For binary systems, the characteristic value D^* equals the diffusion coefficient D. For systems with $n_c > 2$, the geometric mean of the eigenvalues $D_i^\dagger$ of the diffusion coefficient matrix $\boldsymbol{D}$

$$D^* = \left(\prod_{i=1}^{n_c-1} D_i^\dagger\right)^{1/(n_c-1)} \tag{3.5}$$

is used. Small Fourier numbers $Fo \to 0$ correspond to negligible diffusional equilibration. Large Fourier numbers correspond to almost equilibrated concentrations.

3.4 Model-Based Experimental Procedure

The developed experimental procedure is based on the convection-diffusion model (Sec. 3.3) and model-based optimal experimental design from previous work by Bardow *et al.*[36,73] The model-based optimal experimental design affects the design of the microfluidic chip, the composition of the inlet solutions and the contact time of the solutions in the microchannel.

The assumptions in the convection-diffusion model limit the Reynolds number Re to be of order 1 and the Péclet number to $Pe > 100$. The entrance flow region and the thermal entry length have to be negligible ($L_f \sim 0$, $L_e \sim 0$), as well as the reorientation of the interface ($\zeta(s) \sim 0$). To avoid secondary flows induced by the spiral, the restrictions $De^2 Sc \ll 100$, $C/(W/2) \gg 4$, and $De < 1$ are necessary. As the ratio of channel width to channel height is $W/H > 8$, diffusion is much faster in z-direction and therefore concentration differences are negligible.[150]

Additionally, the contact time is optimized by model-based optimal experimental design techniques to increase the measurement sensitivity: The optimal Fourier number Fo_{opt} is in the range from 0.30 to 0.39 for channel geometries $0.05 < H/W < 1$ as shown by Häusler *et al.*[36] For the presented microchannel (Sec. 3.2), the optimal Fourier number is $Fo_{opt} = 0.37$ based on equation (2.43). At this optimal Fourier number, there is considerable progress of diffusional mixing across the whole channel width while the concentration change over the channel width is still observable.

The dimensionless numbers Re, Pe, Fo, De, Sc and the parameters L_f, L_e, $\zeta(s)$ comprise substance properties (density, viscosity, thermal conductivity and diffusion coefficient), dimensions of the microchannel, and the contact time. The dimensions of the microchannel and the volume flow have been chosen to allow for high accuracy and precision of the measured diffusion coefficients because the substance properties are inherent to the system to be measured. The following dimensionless numbers and characteristics apply on experiments in this work (Table 3.4 based on Sec. 2.6.3): Re ranges from 3 to 9.5 (< 10, laminar flow), $Pe =$ from 3478 to 30 942 ($\gg 1000$, dominant convection in s-direction), $De^2 Sc$ from 9 to 83 (< 100, no secondary flows), and Fo from 0.11 to 1 (~ 0.37, suitable contact time). All fluids have the desired temperature after $L_e < 81\,\mu m$, *i.e.*, within the channel inlets. The flow is fully developed after $L_f = 49\,\mu m$ to $64\,\mu m$ in the main channel, and steady state is established after τ_e =0.11 s to 0.33 s after a change. The inversion of the interface would be $\zeta(s = 0.69\,\mathrm{m})$ =39 µm to 273 µm without diffusion but is not of practical importance as diffusion decreases the driving density difference and, therefore, the inversion width. Thus, the chosen dimensions of the presented microchannel (Sec. 3.2) allow for a wide range of substance properties regarding the characteristics of the diffusion experiment.

The microfluidic glass chip provides *in situ* access to the diffusion process by non-invasive Raman microspectroscopy. Four observation points s_1 to s_4 are chosen (illustrated in Fig. 3.1C), Table 3.5) to reduce experimental effort, *i.e.*, the observation points are on a straight line undisturbed by the microfluidic chip holder.

These observation points also make it possible to observe the diffusion process at several contact times, *i.e.*, various Fourier numbers *Fo*. At each observation point, Raman spectra are acquired every 2.5 µm along the channel cross-section y (Fig. 3.1B)). For every system, an exposure time for the spectra acquisition is chosen once to ensure a sufficient signal-to-noise ratio. The exposure times for all systems are listed in Table 3.6. Exposure times were chosen in the range from 1 s to 7 s. Since the experiment is in steady state, acquisition time is not an issue.

As shown by optimal experimental design theory by Bardow *et al.*, for ternary diffusion coefficients, the accuracy of the estimated diffusion coefficient is significantly improved by an optimal independent second experiment.[32] Because 1D Raman diffusion experiments, microfluidic diffusion experiments also profit from independent experiments. Both independent ternary experiments have the same average concentration but should differ maximally in the composition of the inlet solutions, as shown in Fig. 3.2 A). For quaternary diffusion alike, three independent experiments were performed, see Fig. 3.2 B). Thus, $n_c - 1$ independent experiments were performed for a multicomponent diffusion coefficient. Each independent experiment was repeated at least three times. In the following, we use the term series for the repetitions of one independent experiment. Hence, two series of experiments were performed for ternary systems and three series for the quaternary system.

For every experiment, solutions of compositions $\boldsymbol{x}_l$ and $\boldsymbol{x}_r$ are prepared by weight. The sample preparation can result in uncertainties of up to 0.2 mol % due to the uncertainty of the balance (0.3 mg) and the vapor pressures of the solvents. In this work, both solutions contain all components i. Mole fraction differences of components 1 to

Table 3.4: Dimensionless quantities and characteristics of microfluidic diffusion experiments. The mole fraction of the n_c-th component is $x_{n_c} = 1 - \sum_{i=1}^{n_c-1}$. Re is calculated based on equation (2.26), Pe on (2.37), Fo on (2.42), De on (2.33), Sc on (2.34), the inversion ζ on (2.36), the length of the entrance flow region L_f on (2.27), the time to steady state τ_e on (2.30) and the pressure drop Δp over the channel length of 1 m based on (2.32).

	x_1	x_2 in mol %	x_3	$\bar{v}$ in mm s^{-1}	Re	Pe in 10^3	Fo at s_1	Fo at s_4	De	$De^2 Sc$	ζ	L_f in µm	τ_e in s	Δp in bar
cyclohexane (1) + toluene	60			31.3	3.0	3.5	0.002	1.00	0.15	9.0	115	49	0.21	1.4
acetone (1) + toluene	40			62.5	8.6	4.6	0.002	0.75	0.42	33.9	105	66	0.30	2.1
	59			62.5	9.5	4.2	0.002	0.82	0.46	34.2	105	64	0.33	1.9
cyclohexane (1) + methanol	1			62.5	6.6	4.8	0.002	0.72	0.32	26.9	51	56	0.23	2.6
	5			62.5	6.4	6.8	0.001	0.51	0.31	37.0	39	55	0.23	2.6
acetone (1) + water	19.2			62.5	3.1	18	0.000	0.19	0.15	48.6	63	49	0.11	6.3
1-propanol (1) + 1-chlorobutane (2) + heptane	33.3	33.3		62.5	6.7	7.7	0.001	0.45	0.33	44.5	273	56	0.24	2.4
cyclohexane (1) + toluene (2) + methanol	5	5		62.5	6.3	6.6	0.001	0.52	0.31	35.9	68	55	0.22	2.7
	10	10		62.5	6.2	8.4	0.001	0.41	0.30	44.5	87	55	0.22	2.7
water (1) + acetone (2) + toluene	85.17	14.74		62.5	3.1	31	0.000	0.11	0.15	83.0	46	49	0.11	6.4
	80.79	19.07		62.5	3.2	16	0.000	0.21	0.16	44.8	68	49	0.11	6.2
	70.28	29.07		62.5	3.8	16	0.000	0.22	0.18	49.9	46	50	0.13	5.1
	62.95	35.6		62.5	4.3	15	0.000	0.23	0.21	56.0	104	51	0.15	4.3
cyclohexane (1) + toluene (2) + acetone (3) + methanol	5	5	5	62.5	6.8	5.3	0.001	0.65	0.33	30.9	102	56	0.24	2.5

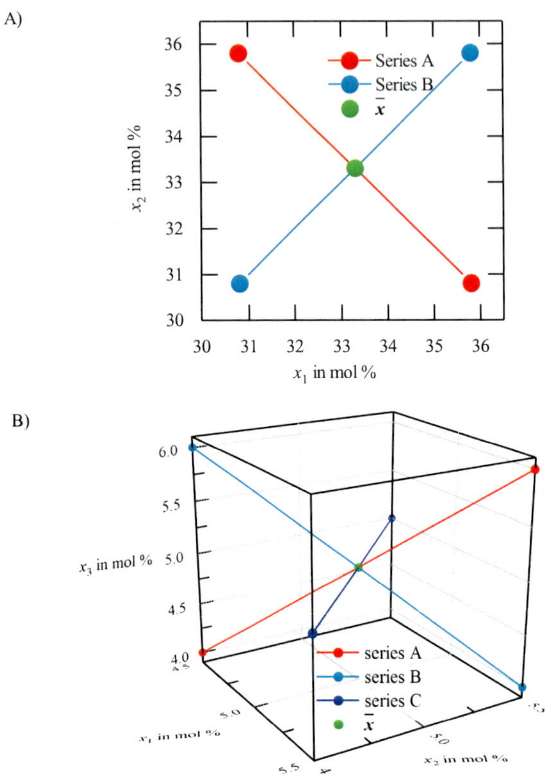

Figure 3.2: $n_c - 1$ series of experiments are performed at the same average composition $\bar{\boldsymbol{x}}$ to determine a multicomponent diffusion coefficient $\boldsymbol{D}^V(\bar{x})$. The n_c-th mole fraction results from $\sum x_i = 1$. (A) Example of two series of independent experiments for 1-propanol (1) + 1-chlorobutane (2) + heptane with compositions listed in Table H.1. (B) Three series of independent experiments for cyclohexane (1) + toluene (2) + acetone (3) + methanol with compositions given in Table K.1.

Table 3.5: Distances of the observation points s_1 to s_4 to the position where the two channel inlets join

	distance in mm
s_1	14.3
s_2	225.8
s_3	452.4
s_4	694.0

$n_c - 1$ are balanced by the n_c-th component. For each system, the inlet compositions and average compositions are listed in the appendix (Table 1 in Sections E- K). The mole fraction difference $\boldsymbol{\Delta x}$ between the inlet solutions is chosen as small as possible to ensure negligible excess volume effects, constant viscosities and constant diffusion coefficients. Simultaneously, $\boldsymbol{\Delta x}$ is chosen larger than the measurement uncertainty of the mole fractions, *i.e.*, sufficiently to distinguish the mixture spectra. To illustrate the difference between the mixture spectra, typical spectra for inlet solutions of the system 1-propanol + 1-chlorobutane + heptane are shown in Fig. 3.3. Despite the small concentration differences and overlapping spectral bands, the spectra are still distinguishable for the eye.

For the validation of the measurement method, three binary and two ternary systems are investigated, see Sec. 3.1. Additionally, new diffusion coefficients were determined for the binary system cyclohexane + methanol, the ternary system water + acetone + toluene and the quaternary system cyclohexane + toluene + acetone + methanol, see Sec. 3.1.

All diffusion experiments are conducted isothermally at temperatures of 25.0(5) °C, as the microfluidic chip is temperature controlled and the thermal entrance length is shorter than the channel inlets. For a diffusion experiment, the inlet solutions are pumped through the channel inlets into the microfluidic chip. The volume flow is chosen once per system with $\dot{V} = 2\dot{V}_\text{l} = 2\dot{V}_\text{r}$ (Table 3.6). The volume flow $\dot{V}$ is adjusted to ensure Fourier numbers from $Fo = 0.00$ at the observation point s_1 up to $Fo = 0.11$ for water + acetone + toluene and $Fo = 1.00$ for cyclohexane + toluene

Table 3.6: Experimental parameters of the microfluidic diffusion experiments

	x_1	x_2	x_3	$\dot{V}$ in	$t_{exposure}$	ν_{start}	ν_{end}
	in mol %			mL min^{-1}	in s	in cm^{-1}	
cyclohexane (1) + toluene	60			0.03	1	500	1800
acetone (1) + toluene	40			0.06	1	2600	3916
	59			0.06	1	2600	3916
acetone (1) + water	19.2			0.06		2600	3916
cyclohexane (1) + methanol	1			0.06	5	500	1600
	5			0.06	5	500	1600
1-propanol (1) + 1-chlorobutane (2) + heptane	33.3	33.3		0.06	3	500	1600
cyclohexane (1) + toluene (2) + methanol	5	5			2	500	1800
	10	10			2	500	1800
water (1) + acetone (2) + toluene	85.17	14.74		0.06	7	2600	3916
	80.79	19.07		0.06	7	2600	3916
	70.28	29.07		0.06	5	2600	3916
	62.95	35.60		0.06	5	2600	3916
cyclohexane (1) + toluene (2) + acetone (3) + methanol	5	5	5	0.06	3	500	1800

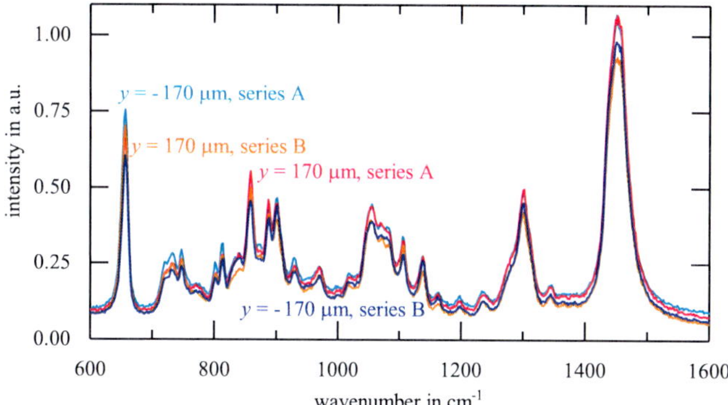

Figure 3.3: Spectra of 1-propanol + 1-chlorobutane + heptane mixtures at the left (—, $y = -170\,\mu\text{m}$, $x_{\text{1-propanol}} = x_{\text{1-chlorobutane}} = 35.8\,\text{mol}\,\%$) and the right (—, $y = 170\,\mu\text{m}$, $x_{\text{1-propanol}} = x_{\text{1-chlorobutane}} = 30.8\,\text{mol}\,\%$) channel inlet for series A, and at the left (—, $y = -170\,\mu\text{m}$, $x_{\text{1-propanol}} = 35.8\,\text{mol}\,\%$ and $x_{\text{1-chlorobutane}} = 30.8\,\text{mol}\,\%$) and the right (—, $y = 170\,\mu\text{m}$, $x_{\text{1-propanol}} = 30.8\,\text{mol}\,\%$ and $x_{\text{1-chlorobutane}} = 35.8\,\text{mol}\,\%$) channel inlet for series B.

at s_4 (Table 3.4). This range includes the optimal Fourier number $Fo_{opt} = 0.37$ for the presented microchannel geometry, with exception of the aqueous systems, see Table 3.4. For acetone + water and water + acetone + toluene, smaller Fourier numbers (Fo =0.11 to 0.19$\gg$ 0) were accepted to increase Pe.

At the beginning of every experiment, a start-up period accounts for the time to reach steady state and the effusion of air bubbles from connecting parts through the microchannel. The maximum start-up period is 15 min, which was determined from repeated measurements. This start-up time also covers the time to steady state estimated from eq. (2.30), which was τ_e <0.21 s for all systems. The overall sample consumption is less than 5 mL, while the overall time for an experiment is less than 1 h.

Densities and viscosities of the mixtures were measured at 298.15 K to determine partial molar volumes and to calculate the characteristics and dimensionless numbers of the experiments. Additional viscosity measurements were performed at 297.15 K and 299.15 K for estimations of the temperature-based uncertainty of the diffusion coefficient based on the temperature-dependence correlation in concentrated solutions (eq. (2.22)). The densities of the samples were measured using the oscillating U-tube principle (DMA 4500 M, Anton Paar GmbH). The repeatability of the sample temperature is 0.01 K, the repeatability of the density measurement is 0.000 01 g cm^{-3} for the DMA 4500 M. Viscosities were measured with a rolling-ball viscometer (LOVIS ME, Anton Paar GmbH). The repeatability of the sample temperature is specified as 0.02 K, the repeatability of the viscosity measurement as 0.05 % for the LOVIS ME. The experimental procedure for DMA 4500 M and LOVIS ME was validated measuring the certified standards APN 100 and APN 415 at 298.15 K and 313.15 K.

3.5 Data Analysis and Parameter Estimation

For every experiment, mole fraction profiles are determined from the acquired Raman spectra at the observation points s_1 to s_4. From these mole fraction profiles, the diffusion coefficient $\boldsymbol{D}^V$ is estimated using the convection-diffusion model (3.1)-(3.4).

Raman spectra of liquid mixtures often show nonlinear effects, e.g., changes in the spectral positions or in the shapes of spectral bands.[229] These nonlinear effects are considered by using indirect hard modeling (IHM) to calculate mole fractions from liquid phase Raman spectra.[99] IHM is also suitable for multicomponent systems.[99]

In the IHM analysis, every pure component spectrum is modeled by a sum of peak-shaped pseudo-Voigt functions (component model Φ_i). For all systems, the spectral range used in the IHM analysis is given in Table 3.6. All bands in this spectral range were considered in the component models. Nonlinear effects in the mixture spectra are compensated by degrees of freedom in the component models. The degrees of freedom enable the modeling of spectral shifts of bands by varying the spectral position of the pseudo-Voigt function from ν to $\tilde{\nu}$. The Raman spectrum of the borosilicate glass is an additional component model Ψ used as background. All component models of the system are weighted by factors b_i and added to a mixture model

$$\Omega = \sum_{i=1}^{n_c} b_i \Phi_i + b_{\text{glass}} \Psi \text{ with } b_i \geq 0. \tag{3.6}$$

To calculate the mole fractions from a mixture spectrum, the residuals between the spectrum and the mixture model Ω are minimized in a least-squares procedure. From the mixture model, integrated intensities of the pure components

$$A_i = b_i \int_{\nu_{\min}}^{\nu_{\max}} \Phi_i d\nu \tag{3.7}$$

are obtained for each spectrum. The area is proportional to the amount of substance which allows to determine mole fractions

$$x_{i,\text{meas}} = \frac{A_i}{\sum_{j=1}^{n_c} A_j k_{ij}} \tag{3.8}$$

using calibration factors k_{ij}. The calibration factors k_{ij} were determined in a further least-squares procedure from calibration spectra with known mole fractions x_i:

$$\text{RMSE}_{\text{cal}} = \left(\sum_{c=1}^{n_{\text{spectra}}} \sum_{i=1}^{n_c - 1} \frac{\left(x_{i,\text{meas}} - x_{i,\text{weighed}} \right)^2}{n_{\text{spectra}} \left(n_c - 1 \right)} \right)^{(1/2)} . \tag{3.9}$$

In this work, these calibration spectra were obtained directly from diffusion experiments. For this purpose, the spectra were analyzed at observation point s_1 (see

Fig. 3.1C)) outside the diffusion zone. In this region, the compositions still correspond to the known initial compositions. Thereby, a multitude of calibration spectra is available from diffusion experiments without any additional effort.

The accuracy of concentration measurements close to the channel walls is lower than in the central region.[67] In this region, the fluid signals decreased due to a smaller sample volume, while also the curvature of the channel wall introduced further optical effects. Therefore, the mole fractions measured close to the channel walls are discarded (y =−200 µm to −180 µm and y =180 µm to 200 µm).

The mole fractions $x_{i,\text{meas}}$ are the input to the parameter estimation procedure for the diffusion coefficient matrix $\boldsymbol{D}^V$. The parameter estimation is based on a maximum likelihood procedure. It is assumed that the standard deviations of the mole fractions $\sigma(x_i)$ are independent of the concentration for small mole fraction differences Δx_i. The standard deviations of the mole fractions $\sigma(x_i)$ are specific for each system. Still, the standard deviations appear to be similar for components i within one system in this study based on the $\text{RMSE}_\text{cal}(x_i)$ values determined from calibration. Hence, $\sigma(x_i)$ are assumed to be the same for all components of one system, constant and independent. Therefore, unweighted least squares are used in the parameter estimation of the diffusion coefficient:

$$\min_{\boldsymbol{D}^V,\boldsymbol{\Delta s},\boldsymbol{y}_0,\boldsymbol{x}_\text{l},\boldsymbol{x}_\text{r}} \quad \text{RMSE}_\text{fit} \tag{3.10}$$

s.t.

$$\text{RMSE}_\text{fit} = \left(\frac{\sum_{\mu=1}^{n_\text{exp}} \sum_{i=1}^{n_\text{c}-1} \sum_{o=1}^{n_\text{s}} \sum_{p=1}^{n_\text{R}(o)} \left(x_{\mu,i,o,p,\text{calc}} - x_{\mu,i,o,p,\text{meas}} \right)^2}{n_\text{exp}\,(n_\text{c}-1)\,n_\text{s}\,n_\text{R}\,(o)} \right)^{1/2} \tag{3.11}$$

$$\boldsymbol{D}^V \text{ is positive definite}\,, \tag{3.12}$$

$$-1\,\text{cm} \leq \Delta s_\mu \leq 1\,\text{cm}\,, \tag{3.13}$$

$$-200\,\mu\text{m} \leq y_{0,\mu} \leq 200\,\mu\text{m}\,, \tag{3.14}$$

$$0\,\text{mol}\,\% \leq x_{\text{l},\mu,i}, x_{\text{r},\mu,i} \leq 100\,\text{mol}\,\%\,. \tag{3.15}$$

Here, n_exp denotes the number of experiments, n_c the number of components, n_s

the number of observation points, and $n_{\mathrm{R}}(o)$ the number of Raman spectra along the measurement line at observation points s_1 to s_4. x_{calc} denotes the calculated mole fractions from the convection-diffusion model (3.1)-(3.4) based on the parameter estimation procedure (3.10)-(3.15). The mole fractions x_{meas} are obtained from the diffusion experiments. To account for uncertainties in the initial experimental conditions, several experimental parameters are fitted together with the diffusion coefficient. Δs is an additional offset of the observation point s_1 to the inlet ($l(s_1) = 14.3\,\mathrm{mm} + \Delta s$). One Δs is adjusted for each experiment and is typically in the range of 0 µm to 2 µm. The distances between the observation points s_1 to s_4 are fixed. y_0 adjusts the position where the two channel inlets join. One y_0 is adjusted for each experiment; it is typically in the range of -5 µm to 5 µm. The mole fractions $\boldsymbol{x}_{\mathrm{l}}$ and $\boldsymbol{x}_{\mathrm{r}}$ at the left and right inlet, respectively, are fitted once for each experiment. The diffusion coefficient matrix $\boldsymbol{D}^V$ is simultaneously fitted to n_{exp} experiments. The quality of the fit is described by the root-mean-square error $\mathrm{RMSE}_{\mathrm{fit}}$ (3.11).

The convection-diffusion model (3.1)-(3.4) is solved numerically, using the MATLAB function PDEPE (MathWorks, MATLAB R2016a). The parameter estimation procedure uses the MATLAB function lsqnonlin.

For systems, in which the partial molar volumes $\bar{V}_i$ deviate significantly from the pure component molar volumes V_i^0, the partial molar volumes are used in the convection-diffusion model (3.1)-(3.4) to determine the diffusion coefficient matrix $\boldsymbol{D}^V$. The partial molar volume is determined from

$$\bar{V}_i = V - \sum_{\substack{j=1 \\ j \neq i}} x_j \left(\frac{\partial V}{\partial x_j} \right) , \tag{3.16}$$

where the molar volume of a mixture is

$$V = \sum_{i=1}^{n_c} x_i V_i^0 + V^{\mathrm{E}} . \tag{3.17}$$

To model the excess molar volume V^{E} of multicomponent mixtures, a multicomponent expansion to a Redlich-Kister type function

$$V^{\mathrm{E}} = \sum\sum_{i \neq j} x_i x_j (A_{ij} + B_{ij}(x_i - x_j) + C_{ij}(x_i - x_j)^2 + ...) \,, \tag{3.18}$$

is used with the binary interaction parameters A, B, C.[230] In this simplified Redlich-Kister type function, ternary and higher-order terms are ignored. For a binary system with a Redlich-Kister type function for the excess molar volume V^{E}, the partial molar volumes are

$$\bar{V}_1 = V_1^0 + (A + 3B + 5C)x_2^2 + (-4B - 16C)x_2^3 + 12Cx_2^4 \,, \tag{3.19}$$

$$\bar{V}_2 = V_2^0 + (A - 3B + 5C)x_1^2 + (4B - 16C)x_1^3 + 12Cx_1^4 \,. \tag{3.20}$$

The diffusion coefficient matrix $\boldsymbol{D}^V$ is positive definite according to the thermodynamics of irreversible processes,[3] *i.e.*, all eigenvalues are positive definite (2.3). The corresponding ternary constraints are directly incorporated into the fitting procedure by a parametrization θ (3.22) proposed by Bardow *et al.*:[32]

$$\theta_1 = D_1^{\dagger},\ \theta_2 = D_2^{\dagger},\ \theta_3 = D_{11} - D_{22},\ \text{and}\ \theta_4 = D_{12} \,. \tag{3.21}$$

This parametrization avoids infeasible path strategies and violations of constraints, as the constraint $D_i^{\dagger}$ (2.3) is trivially satisfied. The inverse transformation is performed according to

$$D_{11} = \frac{\theta_1 + \theta_2 + \theta_3}{2},\ D_{22} = \frac{\theta_1 + \theta_2 - \theta_3}{2},\ D_{21} = \frac{(\theta_1 - \theta_2)^2 - \theta_3^2}{4\theta_4},\ \text{and}\ D_{12} = \theta_4. \tag{3.22}$$

In practical applications, division by the unknown parameter θ_4 to compute D_{21} does not lead to numerical difficulties in our experience.

For quaternary diffusion coefficients, a similar parametrization of the diffusion coefficient is presented. It includes the eigenvalues $D_i^{V,\dagger}$ of the diffusion coefficient matrix $\boldsymbol{D}$

$$\theta_q = D_i^{\dagger} > 0 \quad \text{for} \quad i, q = 1, ..., 3, \tag{3.23}$$

and the differences

$$\theta_q = D_{11} - D_{ij} \quad \text{for} \quad 4 \leq q \leq 9 \,, \quad 2 \leq i \leq 3 \quad \text{and} \quad 1 \leq j \leq 3 \,. \tag{3.24}$$

Appendix C contains the detailed parametrization and reparametrization (C.1)-(C.8).

For multicomponent diffusion, the diffusion coefficient matrix depends on the order of the components in the estimation procedure.[3] For 1-propanol + 1-chlorobutane + heptane and cyclohexane + toluene + methanol, the order of components in the parameter estimation is taken from the first publication of the measured systems in the literature[1,2] to obtain the same Fick diffusion coefficients. For the other systems, the order of components in the estimation is also the same order as in the notation of the system.

Standard deviations of the diffusion coefficient D_{ij}^V are calculated from repeated experiments and combinations of repeated independent experiments. Here, one estimation of a diffusion coefficient matrix $\boldsymbol{D}^V$ is called evaluation. For every system, there are as many evaluations with $n_{\text{exp}} = 1$ as there are experiments. For ternary and quaternary systems, the number of evaluations with $n_{\text{exp}} = 2$ or $n_{\text{exp}} = 3$ is the number of all combinations of experiments from two or three independent series. For cyclohexane + toluene + acetone + methanol with 5 repeated experiments in series A, 5 in series B, and 6 in series C, this results in 16 evaluations with $n_{\text{exp}} = 1$, 85 evaluations with $n_{\text{exp}} = 2$, 150 evaluations with $n_{\text{exp}} = 3$, and 1 evaluation with $n_{\text{exp}} = 16$. The standard deviation from repeated evaluations of n_{exp} experiments is

$$\sigma(D_{ij}^V) = \sqrt{\frac{\sum\limits_{n_{\text{evaluations}}} (D_{ij,s}^V - \bar{D}_{ij}^V)^2}{n_{\text{evaluations}} - 1}} \,. \tag{3.25}$$

Here, $n_{\text{evaluations}}$ denotes the number of evaluations for n_{exp} and $\bar{D}_{ij}^V$ denotes the arithmetic mean from $n_{\text{evaluations}}$ evaluations. $D_{ij,s}^V$ is the diffusion coefficient estimated in a single evaluation. Relative errors of the individual diffusion coefficient D_{ij}^V

$$u_r(D_{ij}^V) = \sigma(D_{ij}^V)/D^{V,*} \tag{3.26}$$

are related to the geometric mean $D^{V,*}$ (eq. (3.5)) of the diffusion coefficient matrix

$\boldsymbol{D}^V$. The relative error of the diffusion coefficient matrix $\boldsymbol{D}^V$ is averaged over the relative errors $u_r(D_{ij}^V)$ of the individual diffusion coefficient

$$u_r(\boldsymbol{D}^V) = \sum_{i=1}^{n_c-1} \sum_{j=1}^{n_c-1} (u_r(D_{ij}^V))/(n_c - 1)^2 \, . \tag{3.27}$$

Additionally to the systems measured for validation of the method, diffusion coefficients are measured for the binary system cyclohexane + methanol, the ternary system water + acetone + toluene and the quaternary system cyclohexane + toluene + acetone + methanol. Data from literature is not available for these systems, therefore, the measured diffusion coefficients are compared to diffusion coefficients predicted from engineering models (Sec. 2.3). Comparisons between the result from this work and engineering models were performed for all systems, to be able to assess the error of the engineering models for different systems.

The methods and models presented in Sec. 2.3 and the Appendix A.1 are used to predict the Maxwell-Stefan diffusion coefficient in concentrated solutions. For this purpose, diffusion coefficients were taken from literature at infinite dilution as well as pure component self-diffusion coefficients. If these diffusion coefficients were not available, mutual and self-diffusion coefficients at infinite dilution were predicted using the Wilke-Chang equation (2.16). To predict binary Maxwell-Stefan diffusion coefficients in concentrated solutions, the Darken (2.17) and the Vignes equations (A.2) were used. To predict multicomponent Maxwell-Stefan diffusion coefficients $Đ_{ij}$, the predictive Darken-LBV equation (2.18) has been used as well as the Darken-KvB (A.14), the generalized Vignes equation (A.2) with the approaches Vignes-WK (A.3), and Vignes-KT (A.4), Vignes-VKB (A.5), Vignes-DKB (A.6), and Vignes-RS (A.7) on $Đ_{ij}^{x_k \to 1}$, and the Vignes-LBV (A.8). Then, the inverse of the Maxwell-Stefan multicomponent diffusion coefficient $\boldsymbol{B}$ is calculated ((2.11)-(2.12)).

To calculate the Fick diffusion coefficients from the Maxwell-Stefan diffusion coefficients using the relation 2.14 between Fick and Maxwell-Stefan diffusion coefficients, the thermodynamic factor is obtained from PC-SAFT as well as COSMO-RS (COSMOtherm X16). The PC-SAFT parameters are taken from Kleiner and Sadowski[231] as well as Gross *et al.*[232–234] In all cases, the binary interaction parameter is set $k_{ij} = 0$ as the PC-SAFT parameters were not fitted to phase equilibria. For acetone

+ water and for acetone + methanol, also association interactions are considered (Sec. 4.3, 4.7, and 4.8).

To transfer the diffusion coefficients from the molar M to the volume V reference velocity frame, the reference frame transformation (2.5)-(2.7) is applied. For multicomponent diffusion coefficients, the prediction error of the individual diffusion coefficient is

$$\mathrm{error}_{\mathrm{prediction}}(D_{ij}) = \frac{D_{ij,\mathrm{meas}} - D_{ij,\mathrm{predicted}}}{D^{*}} \tag{3.28}$$

and

$$\mathrm{RMSE}_{\mathrm{pred}} = \sqrt{\frac{\sum_{i=1}^{n_c-1}\sum_{j=1}^{n_c-1}(D_{ij,\mathrm{meas}} - D_{ij,\mathrm{predicted}})^2}{(n_c-1)^2}} \tag{3.29}$$

is the RMSE of the prediction of the diffusion coefficient matrix $\boldsymbol{D}^V$.

Chapter 4

Diffusion Coefficients Measured by Raman Microspectroscopy and Microfluidics

Parts of this chapter are adapted from Ref.[17] with permission from The Royal Society of Chemistry:

> C. Peters, L. Wolff, S. Haase, J. Thien, T. Brands, H.-J. Koß, A. Bardow (2017). Multicomponent diffusion coefficients from microfluidics using Raman microspectroscopy. *Lab Chip*, vol. 17, pp. 2768.

Contribution report: Writing the draft, principal author, planning the experimental setup, choosing the chemical systems, planning the experiments, supporting experiments conducted by S. Haase and R. Becka, data evaluation.
Parts of this chapter are adapted with permission from

> C. Peters, J. Thien, L. Wolff, H.-J. Koß, A. Bardow (2019). Quaternary Diffusion Coefficients in Liquids from Microfluidics and Raman Microspectroscopy: Cyclohexane + Toluene + Acetone + Methanol. *J. Chem. Eng. Data*, `https://doi.org/10.1021/acs.jced.9b00632`.[42] Copyright (2019) American Chemical Society.

Contribution report: Writing the draft, principal author, planning the experimental setup, choosing the chemical systems, planning the experiments, supporting experiments conducted by J. Thien and R. Becka, data evaluation.

The experiments in this chapter were performed with assistance by S. Haase, J. Thien, R. Becka, L. Becka, M. Bruckhaus, L. Städtler and L. Reinpold. For all systems, thermodynamic factors based on the PC-SAFT were calculated with support of Sebastian Kaminski.

Isothermal diffusion coefficients have been measured in a microfluidic setup for the binary, ternary and quaternary systems presented in Sec. 3.1. The results include the spectra analysis and the results of the estimation procedure, *i.e.*, fitted mole fraction profiles and the measured diffusion coefficients from eq (3.10). The measured diffusion coefficients are compared to data from the literature and diffusion coefficients predicted from engineering models.

4.1 Binary: Cyclohexane + Toluene

Diffusion coefficients of cyclohexane (1) + toluene have been measured at 298.15 K and $\bar{x}_1 = 60\,\mathrm{mol}\,\%$. The compositions of the solutions used in the experiments are given in Table D.1.

The spectra of cyclohexane and toluene show several distinctive bands in the fingerprint region: cyclohexane at $803\,\mathrm{cm}^{-1}$, $1029\,\mathrm{cm}^{-1}$, $1267\,\mathrm{cm}^{-1}$ and $1445\,\mathrm{cm}^{-1}$, and toluene at $522\,\mathrm{cm}^{-1}$, $787\,\mathrm{cm}^{-1}$, $1005\,\mathrm{cm}^{-1}$, $1032\,\mathrm{cm}^{-1}$ and $1211\,\mathrm{cm}^{-1}$ (Fig. 4.1). Additionally, the integrated areas of the pure components are of similar size. Therefore, a very low error of $\mathrm{RMSE_{cal}} = 0.12\,\mathrm{mol}\,\%$ was achieved from calibration spectra.

For cyclohexane + toluene, the diffusion coefficient $\boldsymbol{D}^V$ and experimental parameters are estimated according to eq. (3.10) from measured mole fractions for each single experiment and for a combination of all experiments.

Fig. 4.2 shows the measured and calculated mole fraction profiles at observation points s_1 to s_4 for a typical binary diffusion experiment. The calculated mole fraction profiles are obtained from the convection-diffusion model (3.1)-(3.4). The diffusion coefficient and the experimental parameters (Δs, y_0, $\boldsymbol{x}_\mathrm{l}$, and $\boldsymbol{x}_\mathrm{r}$) necessary in the convection-diffusion model (3.1)-(3.4) are obtained in the fitting procedure (3.10). As expected, the mole fraction gradient decreases with consecutive observation points s_1 to s_4.

Measured and calculated mole fractions are in excellent agreement for all experi-

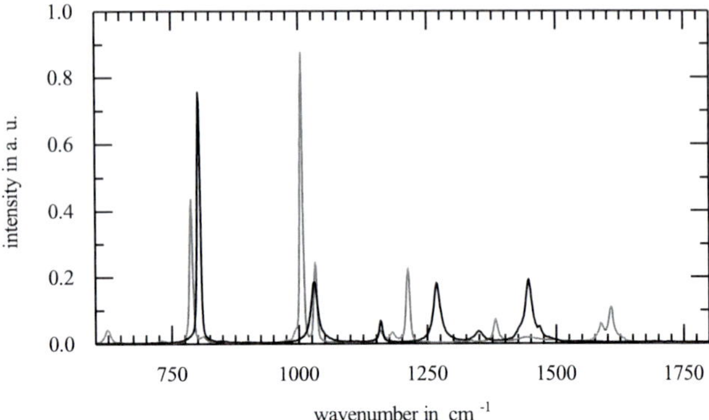

Figure 4.1: Raman spectra of cyclohexane (–) and toluene (–).

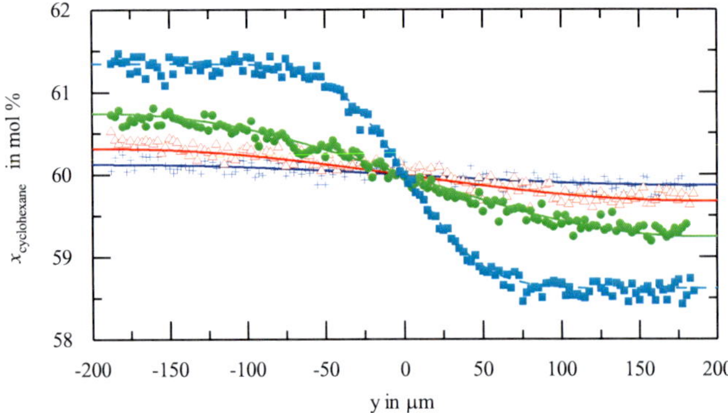

Figure 4.2: Measured (symbols) and fitted (lines, (3.1)-(3.4)) mole fraction profiles of cyclohexane in the system cyclohexane + toluene along measurement line y of the channel cross-sections at observation points $s_1 = 0.0143\,\text{m}$ ($\bar{t} = s/\bar{v} = 0.25\,\text{s}$, ■,–), $s_2 = 0.225\,\text{m}$ ($\bar{t} = 7.41\,\text{s}$, •,–), $s_3 = 0.452\,\text{m}$ ($\bar{t} = 14.89\,\text{s}$, △,–), and $s_4 = 0.694\,\text{m}$ ($\bar{t} = 22.86\,\text{s}$, +,–). The mole fractions of toluene result from $x_{\text{toluene}} = 1 - x_{\text{cyclohexane}}$.

ments, *i.e.*, the individual errors $x_{\text{meas}} - x_{\text{calc}}$ and the RMSE_{fit} =0.079 mol % are similar to the error determined in the calibration RMSE_{cal}=0.12 mol % for V2, see Fig. 4.3. The individual errors are normally distributed with means μ between −0.02 mol % for s_1 and 0.03 mol % for s_3. The standard deviations range from 0.074 mol % to 0.080 mol % for s_1 to s_4. It is assumed that the vectors of spatially resolved errors for s_1 to s_4 are independent. This assumption is supported by correlation coefficients ρ between −0.1 and 0.1 for all combinations of the error vectors of s_1 to s_4.

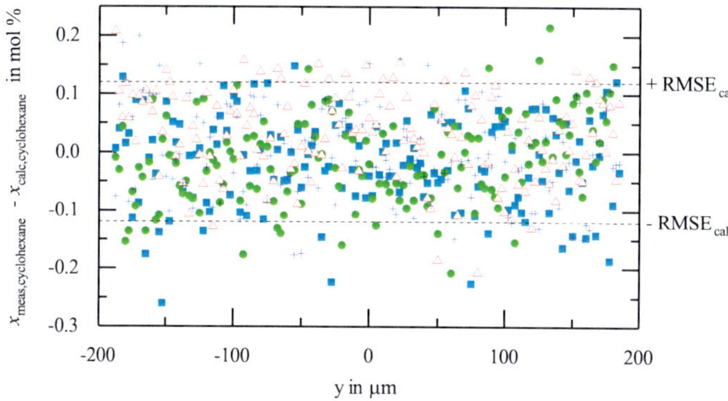

Figure 4.3: Residuals between measured and fitted mole fractions along measurement line y of the channel cross-sections at observation points s_1 (■), s_2 (•), s_3 (△), and s_4 (+) in experiment V2.

The averaged errors for observation points s_1 to s_4 show similar behavior for different experiments, see Fig. 4.4. The average errors are negative for s_1 and s_4 and positive for s_2 and s_3. Still, these errors are smaller than the RMSE_{cal} and not distinguishable within the standard deviation (1σ). Overall, the error between measurement and model is RMSE_{fit}<0.1 mol %. Hence, the RMSE_{fit} is smaller than the RMSE_{cal} of the mole fractions of 0.12 mol % from calibrations.

The good agreement between measurement and convection-diffusion model indicates that the assumptions in the convection-diffusion model are reasonable. Mole fractions measured from Raman y, z-scans in the M.Sc. thesis of Haase[235] show

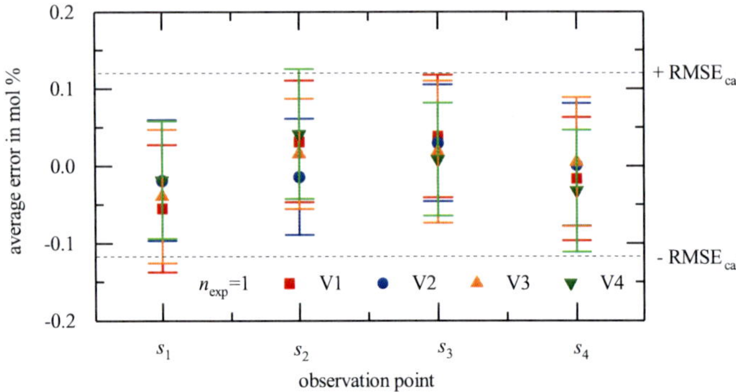

Figure 4.4: Averaged residuals between measured and fitted mole fractions with standard deviations at observation points s_1 to s_4 from four repeated experiments (V1 ■, V2 •, V3 ▲, and V4 ▼) using $n_{\mathrm{exp}} = 1$.

neither the butterfly effect nor reorientation of the interface. y_0 was less than 4 µm in all experiments, although Δy already is 2.5 µm. The small deviations between measurement and model also indicate negligible influences of diffusion in s- and z-direction. For small viscosity differences between the inlet solutions, the inflection point y_0 is approximately at $y = 0$. The curvature of the microfluidic chip is negligible because neither the work in the M.Sc. thesis of Haase[235] nor in the student research project of Juring[236] observed different results when permuting the inlet solutions. All measured mole fractions, along measurement line y and in the cross-section y, z, indicate strictly laminar flow. Potential entrance flow region effects are covered by the estimation parameter Δs.

The average of the diffusion coefficient $\boldsymbol{D}^V$ and its standard deviation $\sigma(\boldsymbol{D}^V)$ are listed in Table 4.1. For the diffusion coefficient $\boldsymbol{D}^V$, a relative error $u_{\mathrm{r}}(\boldsymbol{D}^V) = 2.4\,\%$ was obtained. It is assumed that the main source of uncertainty is the accuracy of the temperature control ($\approx 1\,\%$, eq. (A.19) and Table D.4).

An excellent agreement is found with the measurements of Sanni *et al.*[172] (Table 4.1); The deviation between the diffusion coefficients is 0.45 %. Thus, the diffu-

Table 4.1: Measured diffusion coefficient for the binary system cyclohexane + toluene at $x_{\text{cyclohexane}} = 60\,\text{mol}\,\%$ and $25\,°\text{C}$ from microfluidic experiments in this work with standard deviation $\sigma(\boldsymbol{D}^V)$ from four repeated experiments in comparison to results from Sanni *et al.*[172] For $n_{\text{exp}} = 4$, repeated evaluations are not possible and therefore no standard deviation is available. The quality of fit for the diffusion coefficient is given as RMSE_{fit}.

		$\boldsymbol{D}^V$	$\sigma(\boldsymbol{D}^V)$	RMSE_{fit}
		in $10^{-9}\,\text{m}^2\,\text{s}^{-1}$		in mol %
this work	$n_{\text{exp}} = 1$	1.761	0.039	0.0843
	$n_{\text{exp}} = 4$	1.763		0.0844
Sanni *et al.*[172]		1.767	0.018 to 0.035	

sion coefficient from the microfluidic experiments agrees with the value from Sanni *et al.*[172] within the measurement uncertainty.

Diffusion coefficients were also determined using only 3 of 4 observation points and increasing the step width Δy along the measurement line y, see Fig. 4.5. The number of observation points (three or four) does not influence the uncertainty of the diffusion coefficient but which observation points are omitted does. The uncertainty increases by a factor of 2.5 when omitting s_2 and even by a factor of 5 when omitting s_1. The step width Δy hardly influences uncertainty and precision of the measured diffusion coefficient. The results still agree within the measurement uncertainty, even for $\Delta y = 25\,\mu\text{m}$, using only a total of 60 spectra per experiment. In all these cases, the determined diffusion coefficient still agrees with the diffusion coefficient measured by Sanni *et al.*[172] within the measurement uncertainty.

While results from this work agree well with diffusion coefficients measured by Sanni *et al.*[172] (Fig. 4.6), results by Lin *et al.*[76], Wittko and Köhler[77] and Mialdun and Shevtsova[78] deviate. Diffusion coefficients obtained by Lin *et al.*[76] were averaged over 20 mol %, measured in microfluidics at small Fourier numbers and the minimum at 50 mol % toluene was not observed at 313 K or at 328 K. Wittko and Köhler[77] did not comment on the deviation of up to 12 % to the data of Sanni *et al.*[172]. Mialdun and Shevtsova[78] measured higher diffusion coefficients on ISS (in Fig. 4.6) than on

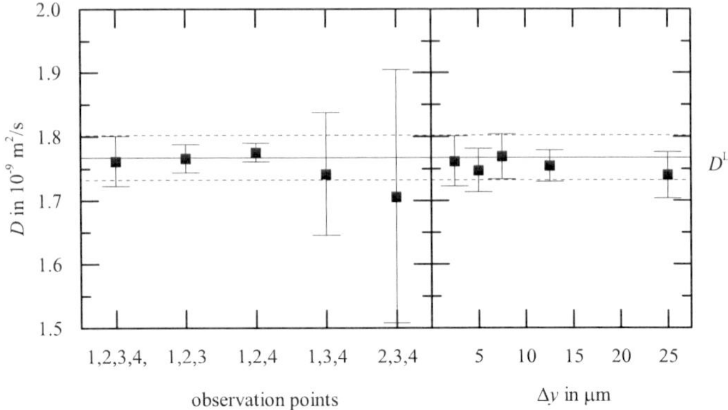

Figure 4.5: Averaged diffusion coefficients with standard deviations from four experiments using $n_{\text{exp}} = 1$, estimated using 4 and 3 observation points as well as using different step width Δy along the measurement line using observation points s_1 to s_4. (–) is the diffusion coefficient measured by Sanni *et al.*, (- -) is a measurement uncertainty of 2 %.

the ground.

Fig. 4.6 also shows diffusion coefficients predicted with engineering models using PC-SAFT and COSMO-RS for the thermodynamic factor. For cyclohexane + toluene, mutual diffusion coefficients at infinite dilution, and self-diffusion coefficients of cyclohexane and toluene are available.[172,237,238] The PC-SAFT parameters were obtained from Gross and Sadowski[232] for nonassociating substances. Using PC-SAFT for cyclohexane + toluene, the Vignes equation (A.1) describes the diffusion coefficients best (deviations from 0 % to 6 %), followed by the Darken-LBV (2.18) and the Darken-KvB (A.14) with deviations up to 10 %. These results are similar to the deviations expected for the Vignes equation 13 % and for the Darken equation 7 % to 11 %.[26,65] Overall, the best agreement between diffusion coefficients from Sanni *et al.*[172] and engineering models is obtained for the toluene rich region. Thermodynamic factors based on COSMO-RS clearly overestimate the non-ideal mixing behavior, resulting in a reversed fit for the engineering models and deviations of up to 30 %. Though, the vapor-liquid equilibrium is qualitatively better predicted by COSMO-RS. Overall, the difference between the engineering models is below 6 %. PC-SAFT is significantly better suited to predict the thermodynamic factor for the diffusion coefficient of cyclohexane + toluene than COSMO-RS.

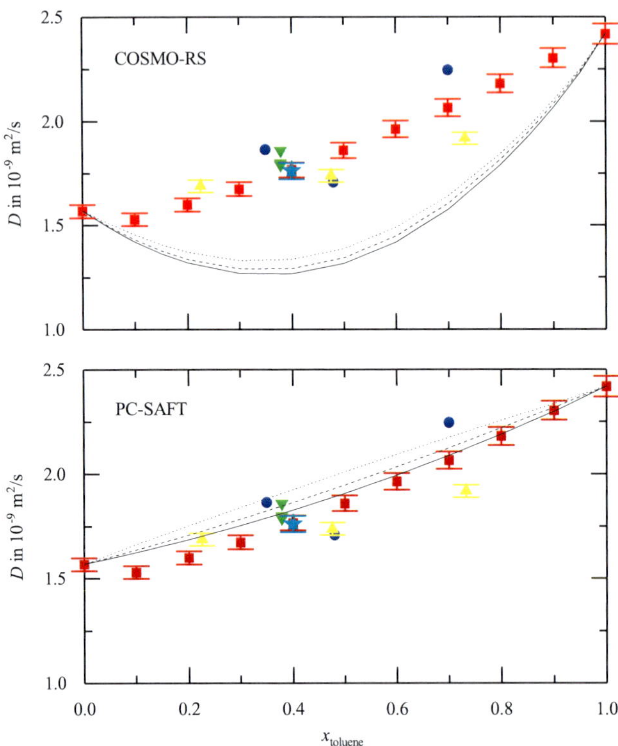

Figure 4.6: Diffusion coefficients of cyclohexane + toluene measured in this work (★) and from literature (symbols) are compared to predictive engineering models (lines). The diffusion coefficients from literature include data from Sanni *et al.*[172] (■), Lin *et al.*[76] (•), Wittko and Köhler[77] (▲) and Mialdun and Shevtsova[78] (▼). Diffusion coefficients are predicted from Vignes eq. (A.1) (–), Darken-LBV (2.18) (- -) and Darken-KvB (A.14) (···) using COSMO-RS (top) and PC-SAFT (bottom).

4.2 Binary: Acetone + Toluene

Diffusion coefficients of acetone (1) + toluene have been measured at 298.15 K and $\bar{x}_1 = \bar{x}_2 = 40.5\,\mathrm{mol}\,\%$ as well as $\bar{x}_1 = \bar{x}_2 = 59.5\,\mathrm{mol}\,\%$. The compositions of the solutions used in the experiments are given in Table E.1.

The spectra of acetone and toluene have been analyzed from $2600\,\mathrm{cm}^{-1}$ to $3916\,\mathrm{cm}^{-1}$. Toluene and acetone show distinctive bands in this spectral range (Fig. 4.7). The integrated area of toluene is 30 % larger than the area of acetone. For acetone + toluene, $\mathrm{RMSE}_{\mathrm{cal}}(x_1 = 40.5\,\mathrm{mol}\,\%) = 0.18\,\mathrm{mol}\,\%$ and $\mathrm{RMSE}_{\mathrm{cal}}(x_1 = 59.5\,\mathrm{mol}\,\%) = 0.3\,\mathrm{mol}\,\%$ are obtained from calibration spectra.

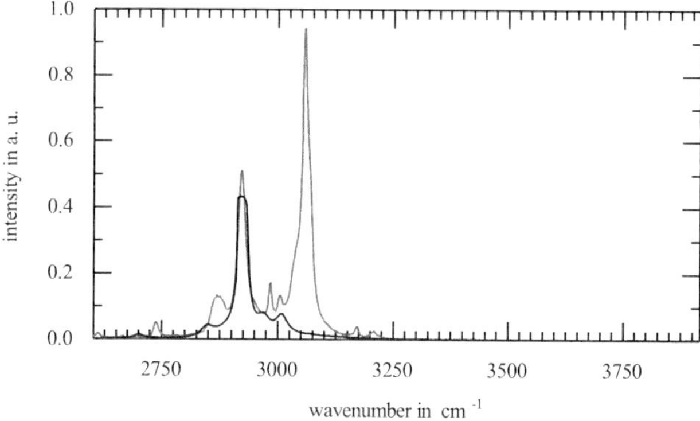

Figure 4.7: Raman spectra of acetone (–) and toluene ( ).

The diffusion coefficient $\boldsymbol{D}^V$ and experimental parameters are estimated according to eq. (3.10) from measured mole fractions for each single experiment and for a combination of all experiments. For both compositions, three experiments were performed.

Fig. 4.8 shows the measured and calculated mole fraction profiles at observation points s_1 to s_4 for a binary diffusion experiment at $x_{\mathrm{acetone}} = 40.5\,\mathrm{mol}\,\%$. Due to the higher diffusion coefficient, the volume flow was increased $\dot{V} = 0.06\,\mathrm{mL\,min}^{-1}$, leading to a continuous decrease of the concentration from s_1 to s_4. The measured and the fitted mole fractions agree well for the observation points s_1 to s_4 for all experiments at both compositions. Also, the $\mathrm{RMSE}_{\mathrm{fit}}$ is similar to the $\mathrm{RMSE}_{\mathrm{cal}}$. Overall, the convection-diffusion model describes the acetone + toluene experiments well.

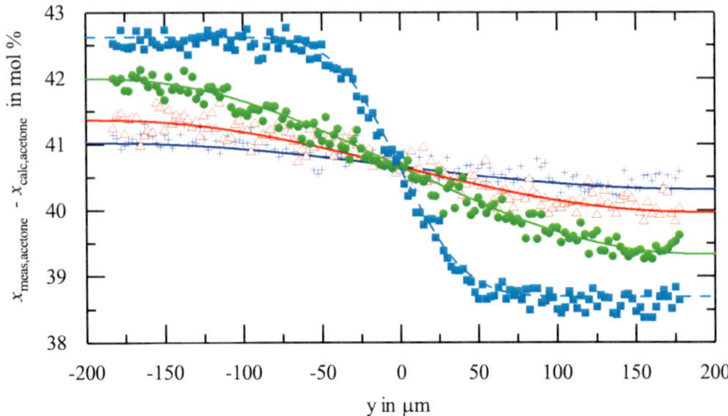

Figure 4.8: Measured (symbols) and fitted (lines) mole fraction profiles of acetone in the system acetone + toluene along measurement line y of the channel cross-sections at observation points s_1 (■,-), s_2 (•,-), s_3 (△,-), and s_4 (+,-).

The diffusion coefficients $\boldsymbol{D}^V$ are given in Table 4.2 for $x_{\mathrm{acetone}} = 40.5\,\mathrm{mol}\,\%$ and $x_{\mathrm{acetone}} = 59.5\,\mathrm{mol}\,\%$. At both compositions, the relative error of the diffusion coefficient is $u_{\mathrm{r}}(\boldsymbol{D}^V) = 4\,\%$.

Measured diffusion coefficients from this work are compared to data from Baldauf and Knapp[212] (Table 4.2 and Fig. 4.9) and Lewis[239]. As Baldauf and Knapp[212] have measured diffusion at 20 °C and 30 °C, the diffusion coefficients were interpolated between these temperatures and extrapolated from 20 °C to 25 °C using the temperature-dependence correlation in concentrated solutions (A.19). The deviation between the interpolated and the predicted diffusion coefficients is 5 %. At 40.5 mol % acetone, the diffusion coefficient agrees with literature data within the measurement uncertainty. At 59.5 mol % acetone, the diffusion coefficient measured in this work

Table 4.2: Measured diffusion coefficient for the binary system acetone + toluene at 25 °C from microfluidic experiments in this work with standard deviation $\sigma(\boldsymbol{D}^V)$ from three repeated experiments in comparison to results from Baldauf and Knapp.[212] The quality of fit for the diffusion coefficient is given as $\mathrm{RMSE_{fit}}$.

		$\boldsymbol{D}^V$	$\sigma(\boldsymbol{D}^V)$	$\mathrm{RMSE_{fit}}$
		in $10^{-9}\,\mathrm{m^2\,s^{-1}}$		in mol %
$x_1 = 40.5\,\mathrm{mol\,\%}$	$n_{\mathrm{exp}} = 1$	2.50	0.09	0.232
	$n_{\mathrm{exp}} = 3$	2.50		0.210
Baldauf and Knapp[212]		2.38	0.08	
$x_1 = 59.5\,\mathrm{mol\,\%}$	$n_{\mathrm{exp}} = 1$	2.90	0.12	0.1588
	$n_{\mathrm{exp}} = 3$	2.86		0.1591
Baldauf and Knapp[212]		2.48	0.09	

deviates 10 % from the diffusion coefficients measured by Baldauf and Knapp.[212] As the fitted and measured concentration profiles look plausible for the later concentration, there was either a systematic error with temperature control or the wrong volume flow in this work. It is also possible that Baldauf and Knapp[212] underestimated the diffusion coefficients of acetone + toluene. This appears plausible as Lewis[239] gives a 15 % larger diffusion coefficient for acetone in toluene at 293.15 K and Tyn and Calus[218] give a larger diffusion coefficient for acetone in water than Baldauf and Knapp. Also, the used chemicals were of different grades, which can cause differing results.

To predict diffusion coefficients of acetone + toluene, self-diffusion coefficients are available.[237,238] For $D_{12}^{x_A \to 1}$, data published by Baldauf and Knapp[212] was interpolated between 293.15 K and 303.15 K. $D_{12}^{x_A \to 0}$ was predicted using the temperature-dependence correlation in concentrated solutions (A.19) based on the diffusion coefficient measured by Lewis.[239] This already shows challenges for engineering models. Even if the required data is available, the quality of the data is unknown which can be seen from the discrepancy between $D_{12}^{x_A \to 0}$ measured by Lewis[239] and measured by Baldauf and Knapp.[212] PC-SAFT parameters were obtained from Gross and Sadowski.[232] Using PC-SAFT to calculate the thermodynamic factor, diffusion

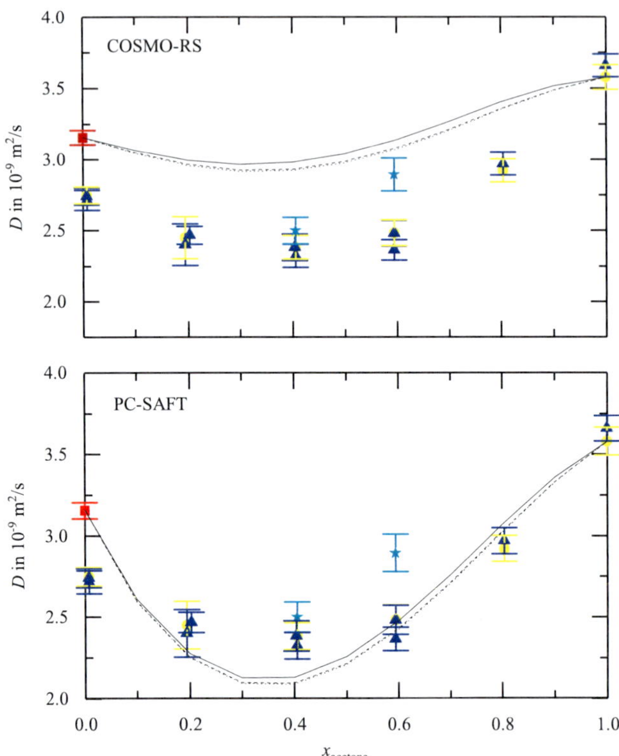

Figure 4.9: Diffusion coefficients of acetone + toluene measured in this work (★) and from literature (symbols) are compared to predictive engineering models (lines). The diffusion coefficients from literature include data from Lewis[239] (extrapolated, ■) and Baldauf and Knapp[212] (interpolated • and extrapolated ▲). Diffusion coefficients are predicted from Vignes eq. (A.1) (–), Darken-LBV (2.18) (- -) and Darken-KvB (A.14) (···) using COSMO-RS (top) and PC-SAFT (bottom).

coefficients are predicted accurately for the acetone-rich region. The diffusion coefficient is underestimated at 40 mol % acetone but was overestimated again for the toluene-rich region (20 mol % acetone). Predicted diffusion coefficients from engineering models together with thermodynamic factors from COSMO-RS overestimate the measured diffusion coefficient by up to 50 %. Also, the VLE is predicted better using activity coefficients from PC-SAFT than from COSMO-RS, though the prediction is good using both models in the acetone-rich region. The difference between Vignes eq. (A.1), Darken-LBV (2.18) and Darken-KvB (A.14) is comparably small for both thermodynamic factors.

4.3 Binary: Acetone + Water

Diffusion coefficients of acetone (1) + water have been measured at 298.15 K and $\bar{x}_1 = \bar{x}_2 = 19.2\,\mathrm{mol\,\%}$. The compositions of the solutions used in the experiments are given in Table F.1.

The spectra of acetone and water have been analyzed from $2600\,\mathrm{cm^{-1}}$ to $3916\,\mathrm{cm^{-1}}$. Water and acetone show distinctive bands in this spectral range (Fig. 4.10). The integrated area of acetone is 3.5 times larger than the area of water. For acetone + water, a $\mathrm{RMSE_{cal}}(x_i) = 0.16\,\mathrm{mol\,\%}$ is obtained from calibration spectra in the diffusion experiments.

For acetone + water, the diffusion coefficient $\boldsymbol{D}^V$ and experimental parameters are estimated according to eq. (3.10) from measured mole fractions for each single experiment and for a combination of all experiments.

Fig. 4.11 shows the measured and calculated mole fraction profiles at observation points s_1 to s_4 for an acetone + water diffusion experiment. As the volume flow was increased for a larger Peclet-number, the Fourier-number decreased to $Fo = 0.19$ and the concentration gradient decreases slower between the observation points s_1 to s_4. Still, measured and calculated mole fractions are in good agreement for all experiments, *i.e.*, the error $\mathrm{RMSE_{fit}} = 0.073\,\mathrm{mol\,\%}$ is even smaller than the error determined in the calibration $\mathrm{RMSE_{cal}} = 0.16\,\mathrm{mol\,\%}$. y_0 was smaller than 12 µm in all experiments. Though a small systematic deviation was observed for s_4 at $y = 150\,\mathrm{\mu m}$ in Fig. 4.11, the standard deviation of $x_{\mathrm{meas}} - x_{\mathrm{calc}}$ is less than 0.08 mol % at all s_1 to s_4 in all three experiments. The mean μ shifts for all three experiments

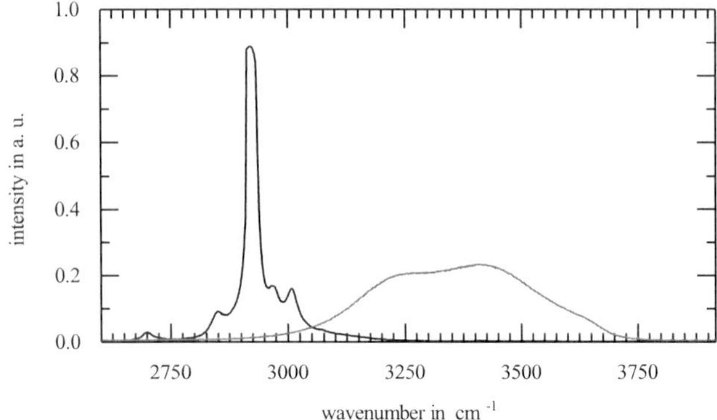

Figure 4.10: Raman spectra of acetone (–) and water (–).

between −0.056 mol % and 0.042 mol % and only a drift for experiment 1. While the correlations are higher between $x_{\text{meas}} - x_{\text{calc}}$ at s_1 to s_4 with r =−0.25 to 0.29 than for cyclohexane + toluene, a linear correlation is still unlikely.

The average of the diffusion coefficient $\boldsymbol{D}^V$ and its standard deviation $\sigma(\boldsymbol{D}^V)$ are written in Table 4.3. For the diffusion coefficient $\boldsymbol{D}^V$, a standard deviation $\sigma(\boldsymbol{D}^V)$ =0.097 $\text{m}^2\,\text{s}^{-1}$ was obtained. The diffusion coefficient determined from experiments in this work agrees well with diffusion coefficient measured by Tyn and Calus:[218] The diffusion coefficients deviate by 7 %, which is within the standard deviation of the diffusion coefficient measured in this work for acetone + water.

Predicting diffusion coefficients using COSMO-RS and engineering models, the diffusion coefficients are accurately described in the water-rich and the acetone-rich region, see Fig. 4.12. For this system, the necessary self and mutual diffusion coefficients are available.[218,238] In the concentrated solution, the diffusion coefficient is overestimated. The change of gradient in the water-rich region for Darken-KvB is due to the contrary gradients of the predicted Maxwell-Stefan diffusion coefficient and the predicted thermodynamic factor based on COSMO-RS. Using PC-SAFT with parameters from Gross *et al.*, $\kappa^{A_iB_j}$=0.034 868 of water for both components, and $\epsilon^{A_iB_j} = 0$ to determine the thermodynamic factor,[231,233,234] the diffusion coefficient

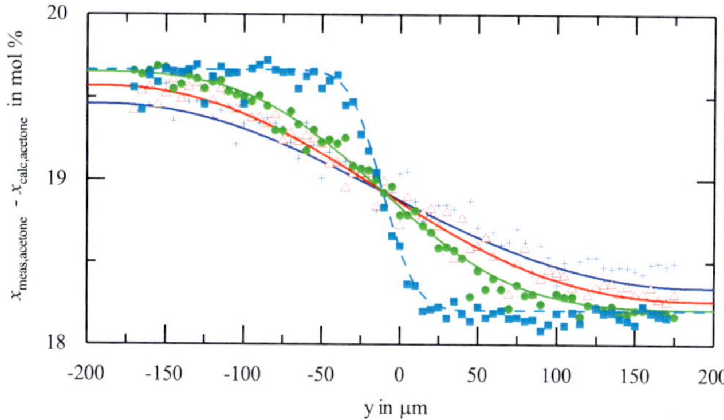

Figure 4.11: Measured (symbols) and fitted (lines) mole fraction profiles of acetone in the system acetone + water along measurement line y of the channel cross-sections at observation points s_1 (■,–), s_2 (•,–), s_3 (△,–), and s_4 (+,–).

Table 4.3: Measured diffusion coefficient for the binary system acetone + water at $x_{\mathrm{acetone}} = 19.1\,\mathrm{mol\,\%}$ and 25 °C from microfluidic experiments in this work with standard deviation $\sigma(\boldsymbol{D}^V)$ from three repeated experiments in comparison to results from Tyn and Calus.[218] The quality of fit for the diffusion coefficient is given as $\mathrm{RMSE}_{\mathrm{fit}}$.

		$\boldsymbol{D}^V$	$\sigma(\boldsymbol{D}^V)$	$\mathrm{RMSE}_{\mathrm{fit}}$
		in $10^{-9}\,\mathrm{m^2\,s^{-1}}$		in mol %
this work	$n_{\mathrm{exp}} = 1$	0.714	0.097	0.073
	$n_{\mathrm{exp}} = 3$	0.700		0.074
Tyn and Calus[218]		0.665	0.01729	

is considerably underestimated. Also, PC-SAFT predicts a non-existent miscibility gap ($D < 0$) and the vapor-liquid equilibrium predicted using PC-SAFT predicts shows less agreement with measured data than the prediction based on COSMO-RS. This may result from the limitations PC-SAFT faces in aqueous systems.[240] Thus, COSMO-RS is better suited to predict the thermodynamic factor for acetone + water. For both PC-SAFT and COSMO-RS, the Vignes equation (A.1) and the Darken-LBV (2.18) describe the diffusion coefficient better than the Darken-KvB (A.14).

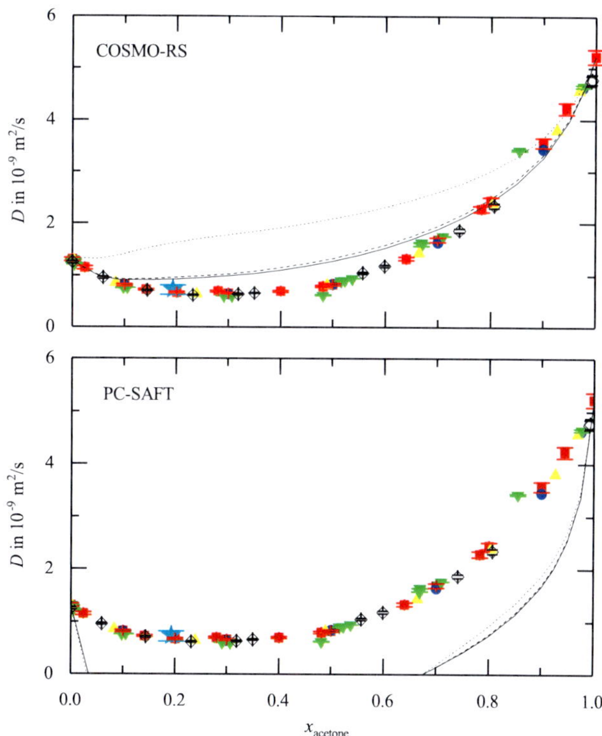

Figure 4.12: Diffusion coefficients of acetone + water measured in this work (★) and from literature (symbols) are compared to predictive engineering models (lines). The diffusion coefficients from literature include data from Tyn and Calus[218](■), Rehfeldt and Stichlmair[241](•), Anderson *et al.*[216] (▲) and Großmann and Winkelmann[242](▼). Diffusion coefficients are predicted from Vignes eq. (A.1) (–), Darken-LBV (2.18) (- -) and Darken-KvB (A.14) (···) using thermodynamic factors determined from COSMO-RS (top) and PC-SAFT (bottom).

4.4 Binary: Cyclohexane + Methanol

Diffusion coefficients of cyclohexane (1) + methanol have been measured at 298.15 K and $\bar{x}_1 = 1\,\text{mol}\,\%$, and $\bar{x}_1 = 5\,\text{mol}\,\%$. A characteristic of this system is the miscibility gap at 298.15 K between 87.6 mol % and 17.7 mol %.[220]

The spectra of cyclohexane and methanol have been analyzed from $500\,\text{cm}^{-1}$ to $1600\,\text{cm}^{-1}$. Cyclohexane and methanol show distinctive bands in this spectral range (Fig. 4.13). The integrated area of cyclohexane is an order of magnitude larger than the area of methanol. For cyclohexane + methanol, a $\text{RMSE}_{\text{cal}}(x_i) = 0.2\,\text{mol}\,\%$ is obtained from calibration spectra in the diffusion experiments.

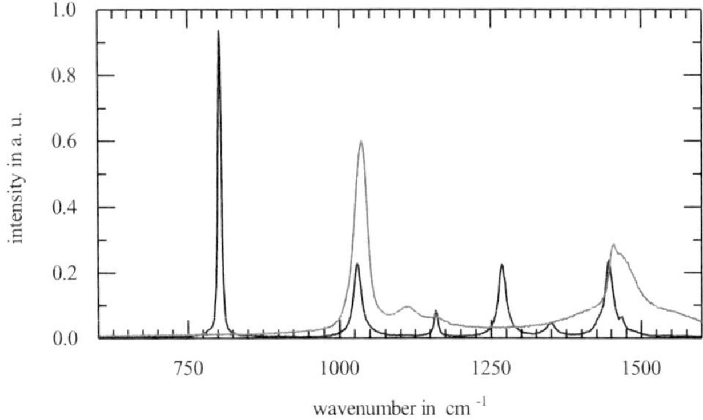

Figure 4.13: Raman spectra of cyclohexane (–) and methanol (–).

For cyclohexane + methanol, the diffusion coefficient $\boldsymbol{D}^V$ and experimental parameters are estimated according to eq. (3.10) from measured mole fractions at each composition for each single experiment and for a combination of all experiments.

Fig. 4.14 shows the measured and calculated mole fraction profiles at observation points s_1 to s_4 for a cyclohexane + methanol diffusion experiment at $x_1 = 1\,\text{mol}\,\%$. The mole fraction gradient decreases visibly, and the mole fraction profiles change significantly from observation point s_1 to s_4. Measured and calculated mole fractions are in good agreement for all experiments, *i.e.*, the error $\text{RMSE}_{\text{fit}} = 0.03\,\text{mol}\,\%$ is almost seven times smaller than the error determined in the calibration $\mathrm{RMSE}_{\mathrm{cal}} = 0.2\,\mathrm{mol}\,\%$. y_0 was smaller than $10\,\mu\mathrm{m}$ in all five experiments.

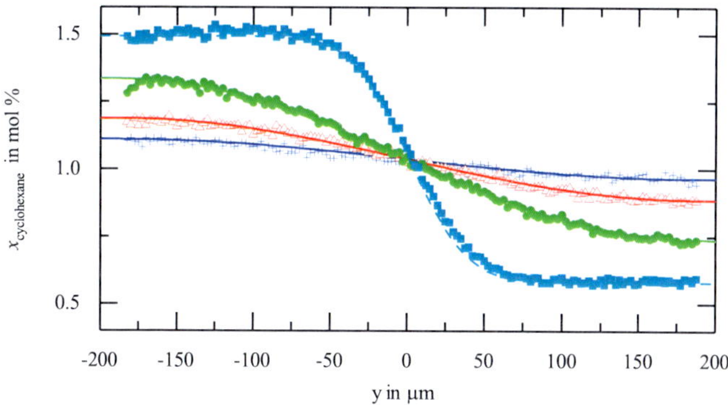

Figure 4.14: Measured (symbols) and fitted (lines) mole fraction profiles of cyclohexane in the system cyclohexane + methanol along measurement line y of the channel cross-sections at observation points s_1 (■,–), s_2 (•,–), s_3 (△,–), and s_4 (+,–) for $x_{\mathrm{cyclohexane}} = 1\,\mathrm{mol}\,\%$.

The average of the diffusion coefficient $\boldsymbol{D}^V$ and its standard deviation $\sigma(\boldsymbol{D}^V)$ are written in Table 4.4 for $x_1 = 1\,\mathrm{mol}\,\%$ and $x_1 = 5\,\mathrm{mol}\,\%$. For the diffusion coefficient $\boldsymbol{D}^V$, relative errors of less than $u_{\mathrm{r}}(\boldsymbol{D}^V) = 5\,\%$ were obtained. The obtained diffusion coefficient at $1\,\mathrm{mol}\,\%$ does not agree well with literature results: Results from this work deviate from results by Tominaga *et al.*[208] by 8 % and from results by Janzen *et al.*.[219] between 7 % and 14 %. The different grades of used chemicals might cause this difference, as Tominaga *et al.*[208] found that diffusion coefficients of cyclohexane differ significantly in alcohols. Cyclohexane of HPLC grade, as used by Tominaga *et al.*, is not unambiguous and Spectronorm only guarantees 99.7 % purity for cyclohexane as used in this work. The methanol quality grades differ also. Other reasons, as systematic errors in this work, by Tominaga *et al.* or by Janzen *et al.* are also possible (see Sec. 4.2). Also, the diffusion coefficient measured by Janzen *et al.* in proximity to the miscibility gap is comparably large. Still, the difference between the diffusion coefficient measured in this work and Tominaga *et al.* has the same order of magnitude as the relative error.

Table 4.4: Measured diffusion coefficient for the binary system cyclohexane + methanol at $x_{\text{cyclohexane}} = 1\,\text{mol}\,\%$ and $x_{\text{cyclohexane}} = 5\,\text{mol}\,\%$ and 25 °C from microfluidic experiments in this work with standard deviation $\sigma(\boldsymbol{D}^V)$ from repeated experiments. The quality of fit for the diffusion coefficient is given as RMSE_{fit}.

		$\boldsymbol{D}^V$	$\sigma(\boldsymbol{D}^V)$	RMSE_{fit}
		in $10^{-9}\,\text{m}^2\,\text{s}^{-1}$		in mol %
this work, 1 mol %	$n_{\text{exp}} = 1$	2.713	0.12	0.019
	$n_{\text{exp}} = 5$	2.685		0.021
this work, 5 mol %	$n_{\text{exp}} = 1$	1.840	0.047	0.041
	$n_{\text{exp}} = 4$	1.837		0.046
Tominaga *et al.*[208], 1 mol %		2.50	0.03	

The diffusion coefficient of cyclohexane + methanol was also predicted using engineering models and the thermodynamic factor from COSMO-RS, PC-SAFT and NRTL (Aspen, APV84 VLE-IG). The necessary self and mutual diffusion coefficients for the engineering models are available,[219,238] though diffusion coefficient of cyclohexane in methanol at infinite dilution was extrapolated from this work. The PC-SAFT parameters were published by Gross and Sadowski.[232,233] While the thermodynamic factor from PC-SAFT predicts complete miscibility, COSMO-RS and NRTL describe a miscibility gap and the general appearance of the vapor-liquid equilibrium. NRTL overestimates the miscibility gap in the methanol-rich region and underestimates the miscibility gap in the cyclohexane-rich region. COSMO-RS overestimates the miscibility gap in both regions. Predicting diffusion coefficients using COSMO-RS or NRTL and engineering models, the diffusion coefficients are well described in the methanol-rich region, see Fig. 4.15. The prediction using NRTL reduces the deviation of the prediction compared to COSMO-RS in the cyclohexane-rich region. In both regions, Vignes, Darken-LBV and Darken-KvB are equally well suited.

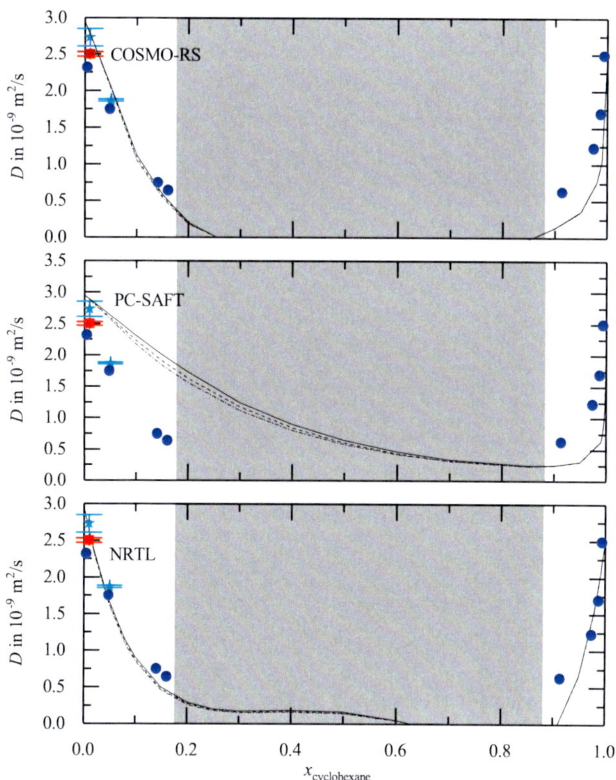

Figure 4.15: Diffusion coefficients of cyclohexane + methanol measured in this work (★) and from literature (symbols) are compared to predictive engineering models (lines). The diffusion coefficients from literature include data from Tominaga *et al.*[208](■) and Janzen *et al.*[219] (•). Diffusion coefficients are predicted from Vignes eq. (A.1) (–), Darken-LBV (2.18) (- -) and Darken-KvB (A.14) (···) using thermodynamic factors determined from COSMO-RS (top), PC-SAFT (middle), and NRTL (bottom). The miscibility gap is indicated by ■.

4.5 Ternary: 1-Propanol + 1-Chlorobutane + Heptane

Diffusion coefficients of 1-propanol (1) + 1-chlorobutane (2) + heptane have been measured at 298.15 K and $\bar{x}_1 = \bar{x}_2 = 33.3\,\mathrm{mol\,\%}$ with three different mole fraction differences $\boldsymbol{\Delta x}$. The compositions of the solutions used in the experiments are given in Table H.1.

In the fingerprint region, all three components have overlapping bands, *e.g.*, from $1400\,\mathrm{cm}^{-1}$ to $1500\,\mathrm{cm}^{-1}$, as shown in Fig 4.16. Therefore, assigning the components in the spectral analysis is more difficult. Still, including these overlapping spectral regions increases the accuracy of the mole fraction analysis. The spectral analysis by IHM considered all spectral bands from $500\,\mathrm{cm}^{-1}$ to $1600\,\mathrm{cm}^{-1}$. The spectra at the channel inlets for series A and series B are shown in Fig. 3.3(A). The shown spectra are typical for the mole fraction range used in the presented experiments. The error of mole fractions for 1-chlorobutane and heptane were strongly correlated using IHM due to the overlapping spectral bands. Therefore, first derivative IHM[243] was used to avoid these systematic errors, as the derivatives of the spectral bands were more distinct. Thereby, the overall accuracy was reduced from $\mathrm{RMSE_{cal}}(x_i)$=0.4 mol % to $\mathrm{RMSE_{cal}}(x_i)$=0.5 mol % as the signal of the derivatives was smaller but a systematic error was avoided.

For the system 1-propanol + 1-chlorobutane + heptane, the diffusion coefficient matrix $\boldsymbol{D}^V$ was estimated according to eq. (3.10) by two procedures: In the first procedure, diffusion coefficients were determined for every individual experiment ($n_\mathrm{exp} = 1$) from series A and series B, e.g., experiment A.1. In the second procedure, diffusion coefficients were determined for all combinations of one experiment from series A with one experiment of series B ($n_\mathrm{exp} = 2$), e.g., A.1 with B.3.

The measured and calculated mole fraction profiles are shown for a typical ternary diffusion experiment in Fig. 4.17. The calculated mole fraction profiles are obtained from the multicomponent convection-diffusion model (3.1)-(3.4) in the fitting procedure (3.10). The plots show excellent agreement between the measured and the calculated mole fractions; For example, the $\mathrm{RMSE_{fit}}$ is on average 0.28 mol % and for all evaluations less than 0.31 mol %. For y_0, values $\leq$12 µm were obtained. The agreement between model and experiment retained the same excellent quality inde-

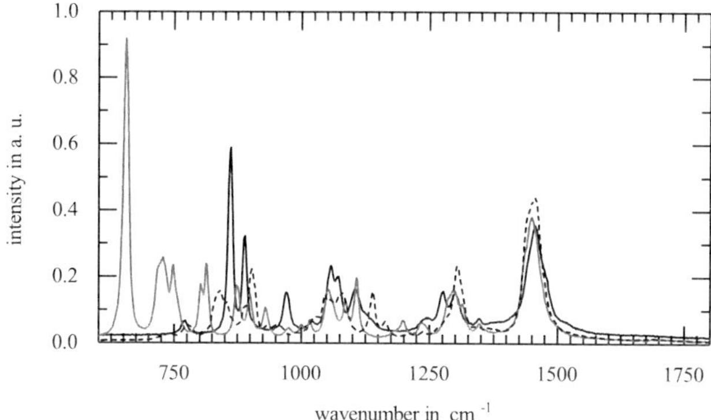

Figure 4.16: Raman spectra of 1-propanol (–), 1-chlorobutane (–), and heptane (- -).

pendently whether the diffusion coefficient was determined from individual experiments or two independent experiments (Table H.7).

Also, the arithmetic average of the $RMSE_{fit}$ is significantly smaller than the $RMSE_{cal}$ for every observation point as shown in Fig. 4.18 for two experiments. Within the standard deviation, the arithmetic averages for all observation points agree. There is no influence of the observation point on the arithmetic mean and the difference between measured and calculated mole fractions are not correlated for the observation points. Hence, the convection-diffusion model is suitable to describe the mass transport for this ternary system.

As shown in Fig. 4.17, the mole fraction gradient of 1-propanol decreases slowest of all three components. The mole fraction gradient of 1-chlorobutane decreases fastest of all three components ($D_{22} > D_{11}$). For heptane, the gradient decreases moderately.

4.5.1 Diffusion Coefficients from Individual Experiments

The estimation of a ternary diffusion coefficient from only one single experiment was successful. The ternary diffusion coefficient matrix $\boldsymbol{D}^V$ from individual experiments

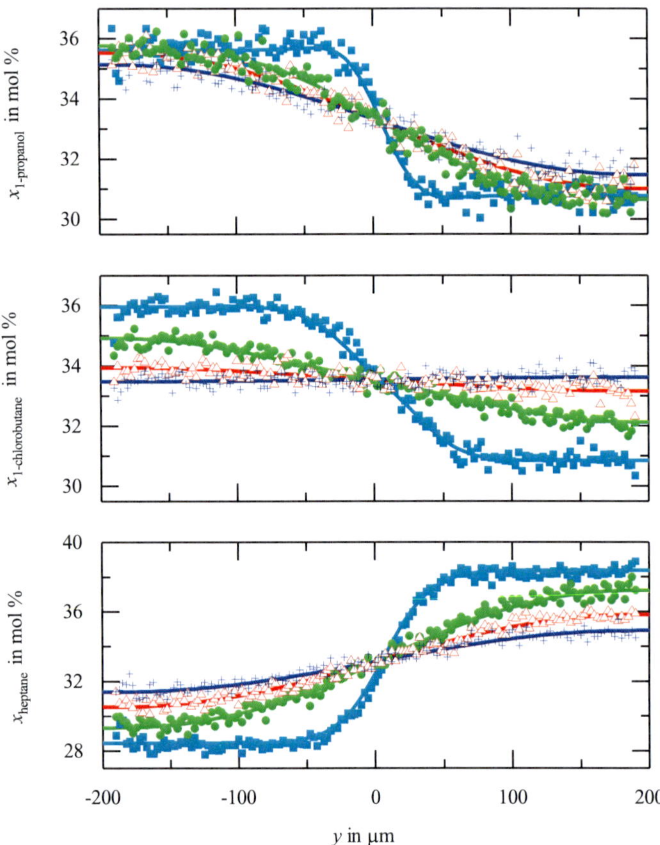

Figure 4.17: Measured (symbols) and fitted (lines) mole fraction profiles of the system 1-propanol + 1-chlorobutane + heptane along measurement line y of the channel cross-sections at observation points $s_1 = 0.0143\,\text{m}$ ($\bar{t} = s/\bar{v} = 0.24\,\text{s}$, ■,–), $s_2 = 0.225\,\text{m}$ ($\bar{t} = 3.71\,\text{s}$, •,–), $s_3 = 0.452\,\text{m}$ ($\bar{t} = 7.44\,\text{s}$, △,–), and $s_4 = 0.694\,\text{m}$ ($\bar{t} = 11.43\,\text{s}$, +,–).

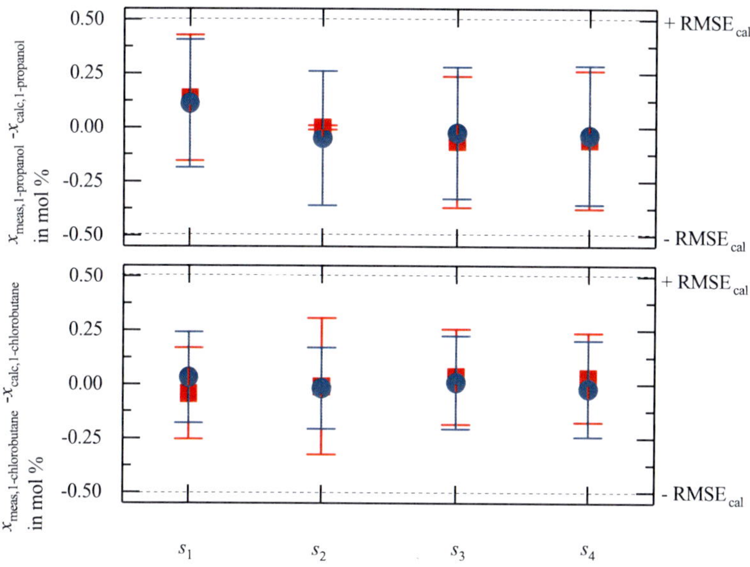

Figure 4.18: Arithmetic mean and standard deviation of the difference between measured and calculated mole fractions for the observation points s_1 to s_4 for 1-propanol and 1-chlorobutane for two independent experiments (■,•).

is shown in the left bars in the matrix of Fig. 4.19 ($n_{\mathrm{exp}} = 1$) for $\Delta x = 5\,\mathrm{mol}\,\%$. The numerical data are given in Table H.7. Although few data was used, reasonable precision and accuracy of the diffusion coefficient $\boldsymbol{D}^V$ were achieved: The relative error of the diffusion coefficient $u_{\mathrm{r}}(\boldsymbol{D}^V)$ is less than 33 %.

Main and cross-diffusion coefficients are correlated as the diffusion behavior cannot be unambiguously assigned to the four diffusion coefficients based on a single experiment. The results agree with expectations as moderate measurement precisions for measurements of multicomponent diffusion coefficients from single experiments are predicted from theory.[32] Notably, the averaged diffusion coefficient matrix agrees with literature data[2] within the measurement uncertainty. Altogether, one microfluidic experiment is sufficient to measure a ternary diffusion coefficient. Determining the multicomponent diffusion coefficient $\boldsymbol{D}^V$ from only one single experiment is only possible with spectroscopic analysis as this analysis provides all concentrations simultaneously. Hence, the experimental effort can be considerably reduced in comparison to classical diffusion experiments.

4.5.2 Diffusion Coefficients from Combinations of Two Independent Experiments

The ternary diffusion coefficient matrix $\boldsymbol{D}^V$ is obtained from combinations of two experiments, one each from series A and series B as shown in the central bars in the matrix of Fig. 4.19 ($n_{\mathrm{exp}} = 2$). In this procedure, parameters were estimated for a total of 16 combinations. From combining 2 experiments, the standard deviations $\sigma(D_{11})$ and $\sigma(D_{21})$ are less than 2 %, while the standard deviations $\sigma(D_{12})$ and $\sigma(D_{22})$ are less than 3.1 %.

The relative errors of the cross-diffusion coefficients D_{12} and D_{21} are similar to the relative errors of the main diffusion coefficients as is typically expected.[244]

The high quality of the fit of the single experiments was retained in the fit of 2 combined experiments ($\mathrm{RMSE_{fit}} = 0.29\,\mathrm{mol}\,\%$, see Table H.7). The second experiment in the evaluation reduces cross-correlations between entries of the diffusion matrix D_{ij}. Thereby, high precision is achieved, and the second independent experiment is highly beneficial. The standard deviation decreases by a factor of 10 in comparison to the estimation from single experiments. It is assumed that the main remaining source

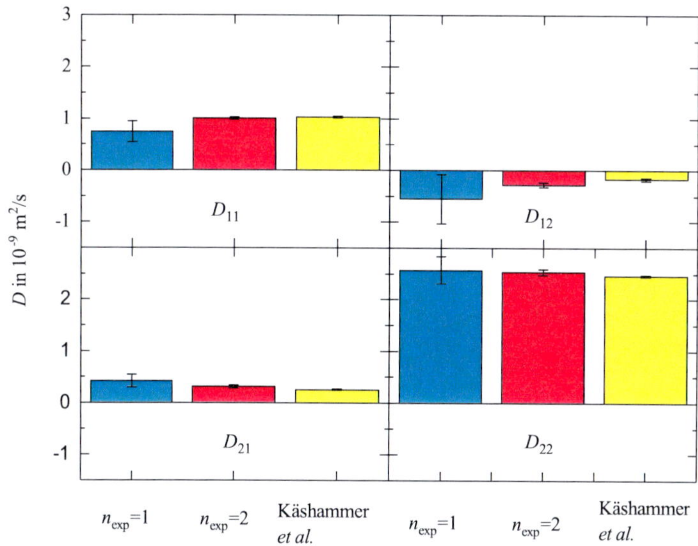

Figure 4.19: Diffusion coefficient matrix $\boldsymbol{D}^V$ with standard deviations of 1-propanol + 1-chlorobutane + heptane at $x_1 = x_2 = 33\,\text{mol}\,\%$ and $T = 25\,°\text{C}$ from this study (■, ■) and from Käshammer *et al.*[2] (■). For $n_{\text{exp}} = 1$ (■), $\boldsymbol{D}^V$ is calculated from a single microfluidic Raman experiment. For $n_{\text{exp}} = 2$ (■), $\boldsymbol{D}^V$ is calculated from two independent experiments.

of uncertainty is the accuracy of the temperature control as in the case of binary diffusion.

A very good agreement with the literature data[2] was achieved. The relative deviations are 2.3 % and 3.4 % for main diffusion coefficients. The main diffusion coefficients agree with the literature values within the measurement uncertainties. Regarding the cross-diffusion coefficients, deviations of less than 8 % from literature were observed. Overall, excellent accuracy, precision, and agreement with literature data were achieved.

As a reference case, all nine available experiments were fitted simultaneously. Even in this case, the quality of the fit stays very high ($\mathrm{RMSE_{fit}} = 0.29\,\mathrm{mol\,\%}$ compared to $\mathrm{RMSE_{fit}}(n_{\mathrm{exp}} = 1) = 0.28\,\mathrm{mol\,\%}$, Table H.7). The diffusion coefficient matrices from $n_{\mathrm{exp}} = 2$ and $n_{\mathrm{exp}} = 9$ agree within the measurement uncertainty of the results from two experiments. Thus, two experiments are sufficient for ternary diffusion experiments in practice.

Additionally, diffusion coefficients were determined with various mole fraction differences between the inlet solutions. Käshammer has already shown that the diffusion coefficient is almost constant for an interval of $\Delta x = 6\,\mathrm{mol\,\%}$ to $10\,\mathrm{mol\,\%}$.[2] This agrees well with the results from this work, see Fig. 4.20. However, by reducing the mole fraction difference, the difference approaches the accuracy of the mole fraction measurements. Therefore, it is noteworthy that the results agree for all three Δx within the measurement uncertainty for $n_{\mathrm{exp}} = 2$, but also the measurement uncertainty is of similar size.

Fig. 4.21 shows the deviations between the measured diffusion coefficient matrix from this work and the predictions from engineering models. For this purpose, diffusion coefficients at infinite dilution were predicted from the Wilke-Chang eq. (2.16), if not available from literature. The multicomponent Maxwell-Stefan diffusion coefficients were predicted from the multicomponent Vignes and Darken extensions presented in eq. (A.2) - (A.14). The Fick diffusion coefficient in the volume reference frame was obtained using eq. (2.13), (2.14) and (2.7).

To predict the diffusion coefficient matrix, four mutual and two self-diffusion coefficients were found in literature,[217,245,246] while the others were predicted with the Wilke-Chang equation (2.16). The PC-SAFT parameters determined by Gross and Sadowski were used in the prediction of the thermodynamic factor.[232,233] Over-

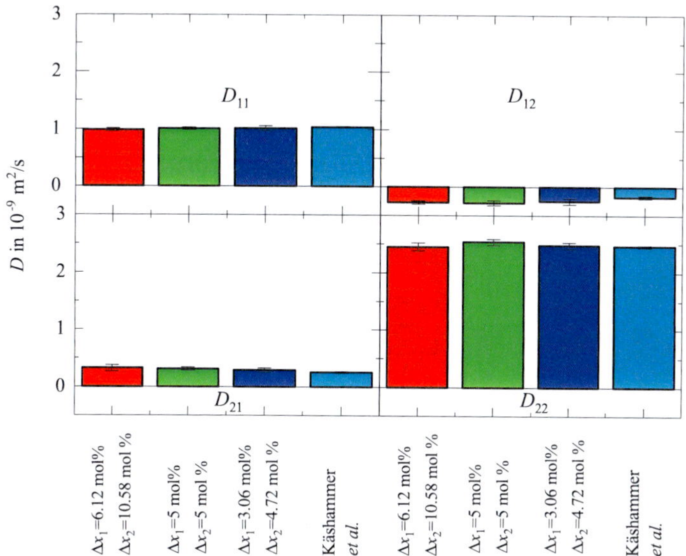

Figure 4.20: Diffusion coefficient matrix $\boldsymbol{D}^V$ with standard deviations of 1-propanol + 1-chlorobutane + heptane at $x_1 = x_2 = 33\,\mathrm{mol}\,\%$ and $T = 25\,°\mathrm{C}$ from this study (■, ■, ■) and from Käshammer *et al.*[2] (■). The diffusion coefficient matrix was determined for $\Delta x_1 = 6.12\,\mathrm{mol}\,\%$, $\Delta x_2 = 10.58\,\mathrm{mol}\,\%$ (■), $\Delta x_1 = 5\,\mathrm{mol}\,\%$, $\Delta x_2 = 5\,\mathrm{mol}\,\%$ (■) and $\Delta x_1 = 3.06\,\mathrm{mol}\,\%$, $\Delta x_2 = 4.72\,\mathrm{mol}\,\%$ (■).

	PC-SAFT		COSMO-RS	
Vignes-WK	-0.298	0.231	-0.519	-0.044
	1.468	0.398	1.173	0.101
	RMSE	0.784	RMSE	0.644
Vignes-KT	-0.156	0.310	-0.448	-0.050
	1.140	0.033	0.914	-0.190
	RMSE	0.596	RMSE	0.518
Vignes-VKB	-0.161	0.304	-0.452	-0.054
	0.642	-0.536	0.515	-0.657
	RMSE	0.452	RMSE	0.475
Vignes-DKB	-0.155	0.310	-0.447	-0.050
	1.176	0.074	0.943	-0.157
	RMSE	0.614	RMSE	0.528
Vignes-RS	-0.228	0.272	-0.483	-0.043
	1.319	0.232	1.055	-0.032
	RMSE	0.693	RMSE	0.581
Vignes-LBV	-0.169	0.284	-0.464	-0.080
	1.258	0.172	1.011	-0.071
	RMSE	0.656	RMSE	0.559
Darken-KvB	0.199	0.587	-0.228	0.068
	1.549	0.500	1.242	0.193
	RMSE	0.871	RMSE	0.640
Darken-LBV	-0.471	0.091	-0.629	-0.109
	0.878	-0.275	0.699	-0.450
	RMSE	0.519	RMSE	0.524

Figure 4.21: Deviations between measured and predicted diffusion coefficient matrix of 1-propanol (1) + 1-chlorobutane (2) + heptane normed by the geometric mean of the measured diffusion coefficient matrix ((3.28), (3.29)). The diffusion coefficient matrices are predicted using eq. (A.2) - (A.14) and the thermodynamic factor (2.13) from PC-SAFT and COSMO-RS. ■ indicates good agreement and ■ large deviations. The $RMSE_{pred}$ considers the individual diffusion coefficients in the matrix.

all, the predictions using the thermodynamic factor from COSMO-RS agree better with the measured diffusion coefficient matrix than the predictions using the thermodynamic factor from PC-SAFT. However, using both predictions, the thermodynamic factor is positive definite. The relative error of the predictions ranges from $\text{error}_{\text{prediction}}(D_{ij})=3\,\%$ to $155\,\%$ for individual diffusion coefficients D_{ij} and $\text{RMSE}_{\text{pred}}$ ranges from $45\,\%$ to $88\,\%$. These results do not agree with expected average errors of $13\,\%$ for the generalized Vignes and even less for the Darken-LBV. Reasons for the large errors are uncertainties in the used diffusion coefficients, the engineering models and the thermodynamic factor. For both thermodynamic factors, the best result was obtained using the Vignes-VKB (A.5).

4.6 Ternary: Cyclohexane + Toluene + Methanol

Diffusion coefficients of cyclohexane (1) + toluene (2) + methanol have been measured at 298.15 K and $\bar{x}_1 = \bar{x}_2 = 5\,\text{mol}\,\%$ as well as $\bar{x}_1 = \bar{x}_2 = 10\,\text{mol}\,\%$. The diffusion coefficient matrix $\boldsymbol{D}^V$ is determined from single experiments as well as from combinations of independent experiments using the estimation procedure (3.10). The compositions of the solutions used in the experiments are given in Table I.1.

The results of the spectra analysis are discussed first, as the Raman spectra are the primary experimental data. Fig. 4.22 shows the pure component spectra of cyclohexane, toluene, methanol, the spectrum of the microfluidic chip and a spectrum of the ternary mixture in the microfluidic chip. The spectra of cyclohexane and toluene show several distinctive bands, *i.e.*, bands of small width with high intensity. The spectrum of methanol shows two distinctive bands at $1037\,\text{cm}^{-1}$ and $1453\,\text{cm}^{-1}$. At the same time, the methanol spectrum provides broad bands of moderate intensity. Hence, cyclohexane and toluene in methanol can be measured with high accuracy even at low quantities. The obtained accuracy of measured mole fractions determined from calibration is $\text{RMSE}_{\text{cal}}(x_i) < 0.3\,\text{mol}\,\%$ for $\bar{x}_1 = x_2 = 5\,\text{mol}\,\%$ and $\bar{x}_1 = \bar{x}_2 = 10\,\text{mol}\,\%$.

All diffusion experiments were performed as described above. For each $\bar{\boldsymbol{x}}$, two series with at least five repetitions were performed. The diffusion coefficients $\boldsymbol{D}^V$ were determined from the measured mole fraction profiles using the least-squares procedure (3.10). The resulting fit between measured mole fractions and fitted mole fractions

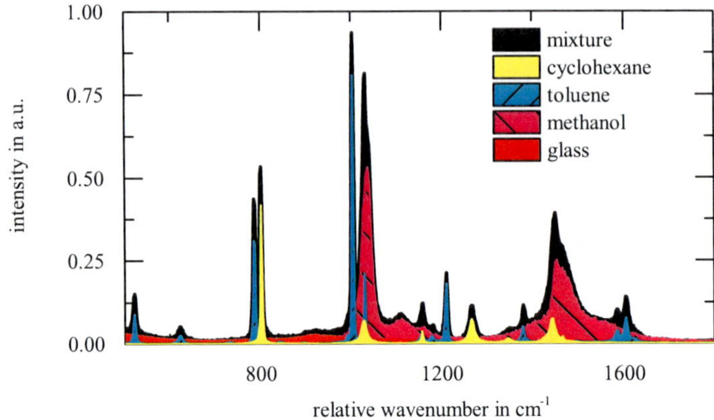

Figure 4.22: Spectra of cyclohexane (1), toluene (2), methanol, glass of the microfluidic chip, and of the ternary mixture ($x_1 = x_2 = 5\,\mathrm{mol}\,\%$) in the microfluidic chip.

from parameter estimation is shown in Fig. 4.23 for $\bar{x}_1 = \bar{x}_2 = 5\,\mathrm{mol}\,\%$ for a typical experiment. Similar excellent agreement between experiment and convection-diffusion model was achieved for all experiments and independently of the number n_{exp} in the evaluation. The $\mathrm{RMSE_{fit}}$ ranges from $0.04\,\mathrm{mol}\,\%$ to $0.06\,\mathrm{mol}\,\%$ for $\bar{x}_1 = \bar{x}_2 = 5\,\mathrm{mol}\,\%$ and is thus an order of magnitude smaller than the calibration error indicating an excellent fit. Excellent agreement between measured and fitted mole fractions (eq. (3.11)) was also achieved for $\bar{x}_1 = \bar{x}_2 = 10\,\mathrm{mol}\,\%$ ($\mathrm{RMSE_{fit}} < 0.08\,\mathrm{mol}\,\%$). y_0 was smaller than 30 µm in all experiments, in 21 of 28 experiments even smaller than 10 µm.

The diffusion coefficient matrix $\boldsymbol{D}^V$ is shown in Fig. 4.24 for $\bar{x}_1 = \bar{x}_2 = 5\,\mathrm{mol}\,\%$, the numerical data are shown in Table I.6. Both Fig. 4.24 and Table I.6 also include diffusion coefficients from the literature. As expected from the concentration gradients of cyclohexane and toluene shown in Fig. 4.23, the main diffusion coefficient of toluene (D_{22}) is larger than the main diffusion coefficient of cyclohexane (D_{11}). Both main diffusion coefficients are significantly larger than the cross-diffusion coefficients.

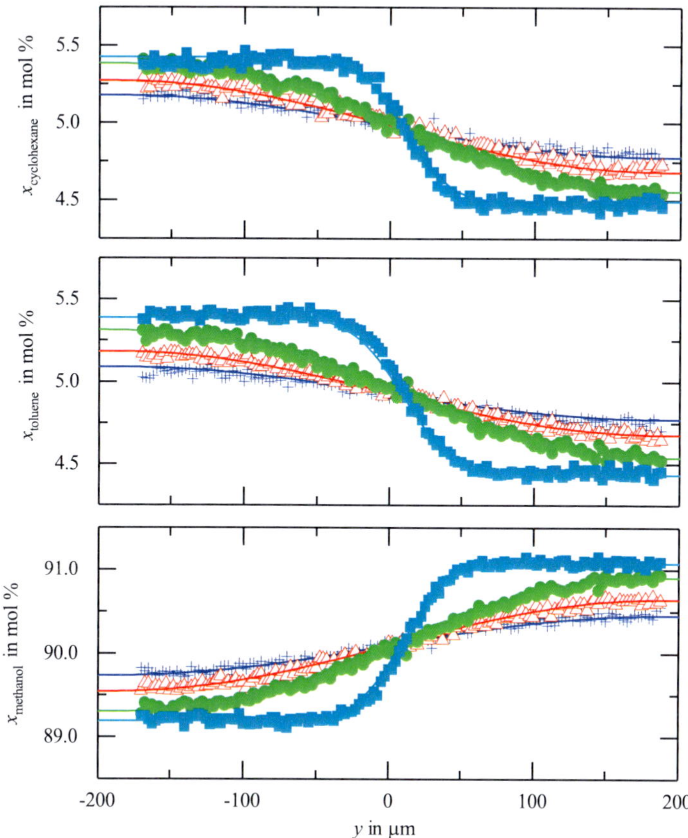

Figure 4.23: Measured (symbols) and fitted (lines) mole fraction profiles of the system cyclohexane (1) + toluene (2) + methanol along measurement line y of the channel cross-sections at observation points s_1 (■), s_2 (•), s_3(▲), and s_4 (+) for $\bar{x}_1 = \bar{x}_2 = 5\,\text{mol}\,\%$. The fit is shown for the determination of the diffusion coefficient from two experiments.

Both cross-diffusion coefficients have a negative sign.

The estimation from single experiments ($n_{\text{exp}} = 1$) is feasible but standard deviations are large. For $n_{\text{exp}} = 1$, the relative error from repeated evaluations is $u_r(\boldsymbol{D}^V) \approx 30\,\%$ to $60\,\%$. The entries of the diffusion coefficient matrix D^V_{ij} are strongly correlated (-0.99 to 0.99). Still, the cross-diffusion coefficients have the correct sign and the averaged diffusion coefficient matrix $\boldsymbol{D}^V$ is very similar to the value determined from combinations of independent experiments. Small relative errors $u_r(\boldsymbol{D}^V) \approx 4\,\%$ are already obtained for the combined analysis of $n_{\text{exp}} = 2$ experiments using one experiment from each series. The standard deviation decreases dramatically from the evaluation of $n_{\text{exp}} = 1$ to the evaluation of $n_{\text{exp}} = 2$, as predicted from optimal experimental design.[32] Therefore, a second independent experiment is worth the effort, as the diffusion coefficient matrix is better accessible.

The diffusion coefficient matrix $\boldsymbol{D}^V$ for $\bar{x}_1 = \bar{x}_2 = 10\,\text{mol}\,\%$ is shown in Fig. 4.25. The corresponding numerical data is given in Table I.7. Both, Fig. 4.25 and Table I.7 also include literature values. The quality and the behavior of the diffusion coefficient matrix are similar to the diffusion coefficient matrix at $x_1 = x_2 = 5\,\text{mol}\,\%$. The measured diffusion coefficient matrices from evaluations with different n_{exp} agree with each other within the measurement uncertainty. The measurement of the diffusion coefficient matrix with $n_{\text{exp}} = 1$ is feasible. In this case, large relative errors (on average $180\,\%$ over D^V_{ij}) are obtained as the entries of the diffusion coefficient matrix are correlated ($r = -0.98$ to 0.99). Still, the average of the diffusion coefficients is similar to the diffusion coefficients determined from combinations of independent experiments, e.g., $n_{\text{exp}} = 2$. Also, the correct sign of the cross-diffusion coefficients is already obtained for $n_{\text{exp}} = 1$. A small relative error $u_r(\boldsymbol{D}^V) = 4\,\%$ is obtained for $n_{\text{exp}} = 2$ using two independent experiments.

The diffusion coefficient matrix $\boldsymbol{D}^V$ was also determined for the combined analysis of all experiments for $\bar{x}_1 = \bar{x}_2 = 5\,\text{mol}\,\%$ and $\bar{x}_1 = \bar{x}_2 = 10\,\text{mol}\,\%$. The results agree with those determined for $n_{\text{exp}} = 1$ and $n_{\text{exp}} = 2$ within the measurement uncertainty. The quality of fit is similar to the case with $n_{\text{exp}} = 2$ (Table I.6, Table I.7). Thus, 2 experiments are sufficient to determine a ternary diffusion coefficient matrix $\boldsymbol{D}^V$.

The diffusion coefficients D^V_{11} and D^V_{21} agree with the results from Großmann and Winkelmann[1] within the measurement uncertainty for $\bar{x}_1 = \bar{x}_2 = 5\,\text{mol}\,\%$ and $\bar{x}_1 = \bar{x}_2 = 10\,\text{mol}\,\%$. In contrast, D^V_{12} and D^V_{22} deviate $3\,\%$ and $15\,\%$ from literature,

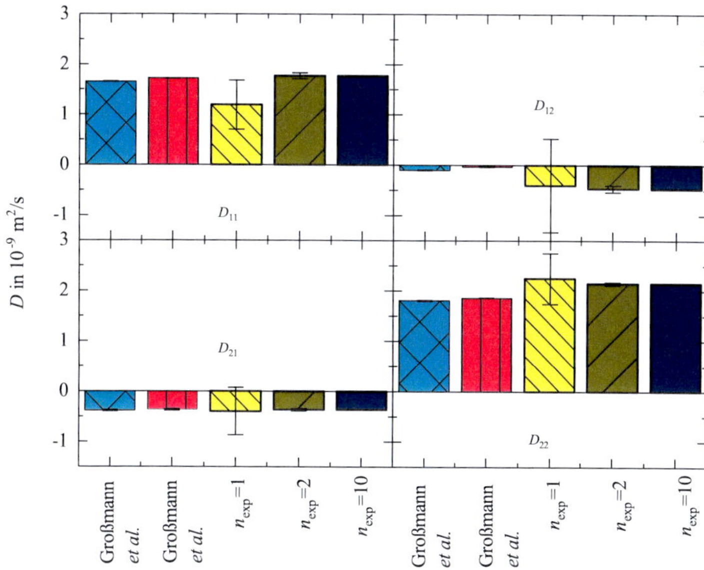

Figure 4.24: Ternary diffusion coefficient matrix $\boldsymbol{D}^V$ of cyclohexane (1) + toluene (2) + methanol at $\bar{x}_1 = \bar{x}_2 = 5\,\mathrm{mol}\,\%$ and $T = 298.15\,\mathrm{K}$ in comparison to Großmann et al.[1] Großmann et al.[1] measured diffusion coefficients at $x_1 = 5.01\,\mathrm{mol}\,\%$ and $x_2 = 4.97\,\mathrm{mol}\,\%$ (■) and at $x_1 = 5.00\,\mathrm{mol}\,\%$ and $x_2 = 5.00\,\mathrm{mol}\,\%$ (■). Results are shown for evaluation of n_{exp} experiments with $n_{\mathrm{exp}} = 1$ (■), $n_{\mathrm{exp}} = 2$ (■) and $n_{\mathrm{exp}} = 10$ (■). Error bars correspond to the standard deviation with a coverage factor $k = 1$. Standard deviations for $n_{\mathrm{exp}} = 1$ ($\sim O1 \times 10^{-9}\,\mathrm{m\,s^{-2}}$) are omitted to allow for clarity. All numerical data are given in Table I.6.

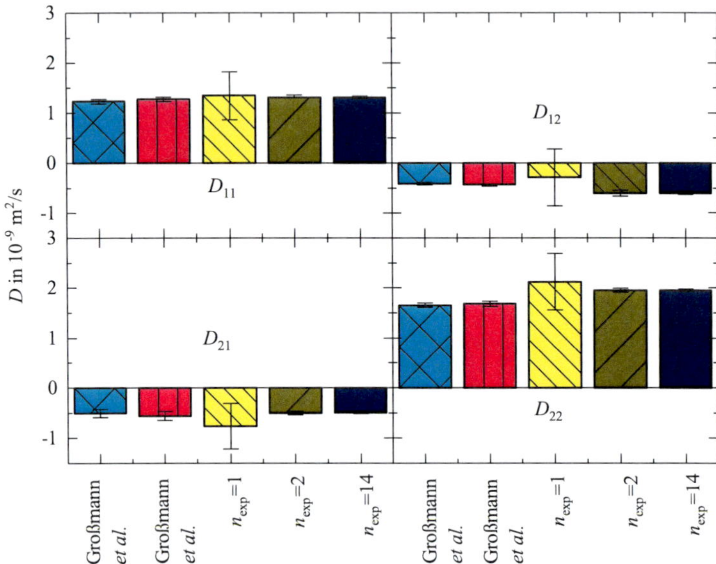

Figure 4.25: Ternary diffusion coefficient matrix $\boldsymbol{D}^V$ of cyclohexane (1) + toluene (2) + methanol at $\bar{x}_1 = \bar{x}_2 = 10\,\text{mol}\,\%$ and $T = 298.15\,\text{K}$ in comparison to Großmann et al.[1] Großmann et al.[1] present two diffusion coefficients $\boldsymbol{D}^V$ for $x_1 = x_2 = 10.00\,\text{mol}\,\%$ (■, ■). Results are shown for evaluation of n_{exp} experiments with $n_{\text{exp}} = 1$ (■), $n_{\text{exp}} = 2$ (■) and $n_{\text{exp}} = 14$ (■). Error bars correspond to the standard deviation with a coverage factor $k = 1$. Standard deviations for $n_{\text{exp}} = 1$ ($\sim O1 \times 10^{-9}\,\text{m}\,\text{s}^{-2}$) are omitted to allow for clarity. Numerical data are given in Table I.7.

respectively.[1] In this work as well as the work by Grossmann and Winkelmann[1], constant diffusion coefficients were assumed. This assumption appears plausible due to the small RMSE_{fit} for $\bar{x}_1 = \bar{x}_2 = 5\,\text{mol}\,\%$ and $\bar{x}_1 = \bar{x}_2 = 10\,\text{mol}\,\%$ in this work. The difference between the results from this work and Grossmann and Winkelmann[1] can originate from different qualities of the used chemicals (HPLC grade used by Grossmann and Winkelmann[1]) and different concentration differences, which were not specified by Grossmann and Winkelmann.[1] Overall, good agreement with literature was still achieved for both compositions.

The diffusion coefficient matrix of cyclohexane + toluene + methanol was also predicted using engineering models presented in Appendix A.1 and COSMO-RS and PC-SAFT. The necessary diffusion coefficients were obtained from this work, the Wilke-Chang eq. (2.16) and literature.[172,237,238,247] The PC-SAFT parameters were obtained from Gross and Sadowski.[232,233] The deviations for the diffusion coefficients and the $\text{RMSE}_{\text{pred}}$ for the matrix between measurements in this work and predictions are shown in Fig. 4.26 and Fig. 4.27. The highest agreement was obtained for both compositions using COSMO-RS together with the Darken-KvB (A.14) and the Darken-LBV (2.18). The lowest accuracy is obtained using the Vignes-VKB (A.5), independent of PC-SAFT and COSMO-RS. Regarding typical errors for predictive engineering models,[26] the difference between Darken-KvB and Darken-LBV and Vignes-VKB is significant, while all other differences have a size similar to the model uncertainties. Generally, the accuracy is higher for the thermodynamic factor determined from COSMO-RS, as also for cyclohexane + methanol. For cyclohexane + toluene + methanol, the thermodynamic factor dominates the $\text{RMSE}_{\text{pred}}$.

	PC-SAFT		COSMO-RS	
Vignes-WK	0.464	0.179	0.184	0.041
	0.139	0.400	0.028	0.288
	RMSE	0.327	RMSE	0.173
Vignes-KT	0.505	0.152	0.219	0.012
	0.100	0.397	-0.002	0.289
	RMSE	0.334	RMSE	0.181
Vignes-VKB	0.522	0.214	0.226	0.067
	0.055	-0.589	0.045	-0.613
	RMSE	0.408	RMSE	0.329
Vignes-DKB	0.492	0.152	0.208	0.014
	0.101	0.384	-0.001	0.277
	RMSE	0.325	RMSE	0.173
Vignes-RS	0.486	0.165	0.202	0.026
	0.118	0.400	0.012	0.290
	RMSE	0.331	RMSE	0.177
Vignes-LBV	0.522	0.155	0.231	0.013
	0.102	0.396	-0.001	0.288
	RMSE	0.340	RMSE	0.185
Darken-KvB	0.435	0.157	0.162	0.023
	0.105	0.311	0.009	0.209
	RMSE	0.283	RMSE	0.133
Darken-LBV	0.471	0.160	0.191	0.022
	0.107	0.339	0.008	0.235
	RMSE	0.306	RMSE	0.152

Figure 4.26: Deviations (3.28) and $\mathrm{RMSE}_{\mathrm{pred}}$ (3.29) between measured ($n_{\mathrm{exp}} = 2$) and predicted diffusion coefficients from engineering models (A.2)-(A.14) for cyclohexane (1) + toluene (2) + methanol at $x_1 = x_2 = 5\,\mathrm{mol\,\%}$ and 298.15 K.

	PC-SAFT		COSMO-RS	
Vignes-WK	0.682	0.188	0.098	-0.110
	0.054	0.591	-0.142	0.356
	RMSE	0.462	RMSE	0.205
Vignes-KT	0.769	0.134	0.159	-0.175
	-0.039	0.585	-0.197	0.371
	RMSE	0.488	RMSE	0.241
Vignes-VKB	0.801	0.254	0.159	-0.080
	-0.112	-0.605	-0.049	-0.619
	RMSE	0.521	RMSE	0.323
Vignes-DKB	0.746	0.135	0.145	-0.169
	-0.038	0.560	-0.192	0.349
	RMSE	0.472	RMSE	0.228
Vignes-RS	0.726	0.159	0.129	-0.144
	0.005	0.591	-0.171	0.366
	RMSE	0.475	RMSE	0.224
Vignes-LBV	0.803	0.141	0.178	-0.176
	-0.037	0.583	-0.196	0.369
	RMSE	0.502	RMSE	0.243
Darken-KvB	0.649	0.151	0.084	-0.134
	-0.021	0.424	-0.160	0.231
	RMSE	0.395	RMSE	0.161
Darken-LBV	0.703	0.159	0.115	-0.139
	-0.014	0.464	-0.162	0.263
	RMSE	0.429	RMSE	0.179

Figure 4.27: Deviations (3.28) and $RMSE_{pred}$ (3.29) between measured ($n_{exp} = 2$) and predicted diffusion coefficients from engineering models (A.2)-(A.14) for cyclohexane (1) + toluene (2) + methanol at $x_1 = x_2 = 10\,mol\,\%$ and 298.15 K.

4.7 Ternary: Water + Acetone + Toluene

Diffusion coefficients of water (1) + acetone (2) + toluene have been measured at 298.15 K in the water-rich region down to $x_1 = 36\,\mathrm{mol\,\%}$. The compositions of the solutions used in the experiments are given in Table J.1.

The spectra of water, acetone and toluene have been analyzed from $2600\,\mathrm{cm}^{-1}$ to $3916\,\mathrm{cm}^{-1}$. Spectra of all three components are also shown in Fig. 4.28. For high water contents, the signal of toluene is limiting the accuracy of the mole fraction measurements, while the overall accuracy is high ($\mathrm{RMSE_{cal}} = 0.07\,\mathrm{mol\,\%}$). The obtained maximal error is $\mathrm{RMSE_{cal}} = 0.22\,\mathrm{mol\,\%}$ at $70.28\,\mathrm{mol\,\%}$ water.

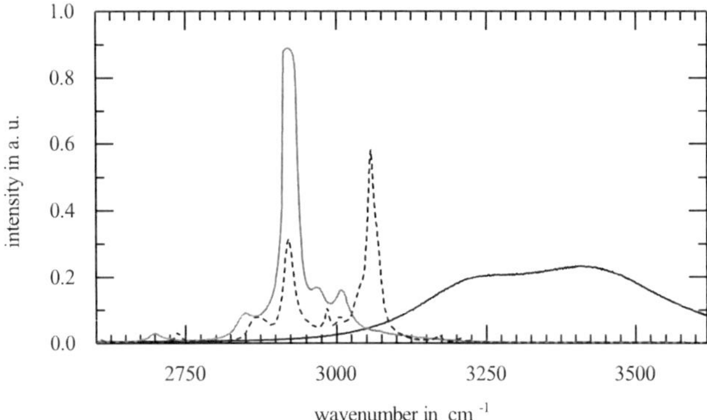

Figure 4.28: Raman spectra of water (–), acetone (–), and toluene (- -).

For water + acetone + toluene, the diffusion coefficient $\boldsymbol{D}^V$ and experimental parameters are estimated according to eq. (3.10) from measured mole fractions using combinations of two experiments as well as all experiments simultaneously. Fig. 4.29 shows the measured and calculated mole fraction profiles at observation points s_1 to s_4 for a water + acetone + toluene diffusion experiment at $x_1 = 62.95\,\mathrm{mol\,\%}$. Measured and calculated mole fractions are in good agreement for all experiments, *i.e.*, the error $\mathrm{RMSE_{fit}} = 0.07\,\mathrm{mol\,\%}$ is similar to the error determined in the calibration $\mathrm{RMSE_{cal}} = 0.07\,\mathrm{mol\,\%}$. For all concentrations, a $y_0 \leq 13\,\mathrm{\mu m}$ was obtained. The mole fraction profiles change significantly; water and acetone even show diffusion against their gradients. The mole fraction gradient of toluene decreases slowly.

The average of the diffusion coefficient $\boldsymbol{D}^V$ and its standard deviation $\sigma(\boldsymbol{D}^V)$ for $n_{\mathrm{exp}} = 2$ are shown in Fig 4.30. The main diffusion coefficient D_{11} is negative for all investigated concentrations and significantly nonzero, *i.e.*, uphill diffusion occurs due to the main and the cross-diffusion coefficients. D_{22} is large, as the acetone concentration gradient decreases fast despite additional uphill diffusion due to cross-diffusion coefficients. The value of D_{12} is significantly larger than the values of all other diffusion coefficients. Still, the diffusion coefficient matrix is positive definite.

For the diffusion coefficient $\boldsymbol{D}^V$, the errors of individual diffusion coefficients range from $0.04 \times 10^{-9}\,\mathrm{m^2\,s^{-1}}$ (D_{21} at x_1 =80.79 mol %) to $6.38 \times 10^{-9}\,\mathrm{m^2\,s^{-1}}$ (D_{12} at x_1 =85.17 mol %). Due to the proximity to the miscibility gap, the geometric mean of the eigenvalues is very small and increases the relative error (up to $u_{\mathrm{r}}(\boldsymbol{D}^V) = 590\,\%$). Related to the individual diffusion coefficients, the relative error is only ~50 %. With increasing geometric mean of the eigenvalues, the relative error decreases to $u_{\mathrm{r}}(\boldsymbol{D}^V)$ =18 %, $u_{\mathrm{r}}(\boldsymbol{D}^V)$ =45 %, and $u_{\mathrm{r}}(\boldsymbol{D}^V)$ =80 % at 80.79 mol %, 70.28 mol %, and 62.95 mol % respectively.

The diffusion coefficient matrix $\boldsymbol{D}^V$ was also predicted using multicomponent Darken and Vignes equations together with thermodynamic factors from COSMO-RS, NRTL, and PC-SAFT. All diffusion coefficients were obtained from literature,[212,218,237,238,248,249], values for acetone + toluene and toluene in water were interpolated. PC-SAFT parameters were obtained from Gross and Sadowski, Gross and Vrabec as well as $\kappa^{A_i B_j}$ =0.034 868 and $\epsilon^{A_i B_j} = 0$ for water and acetone from Klein and Sadowski[231–234] resulting in a non-existent miscibility gap for water + acetone. NRTL parameters were obtained from ASPEN V8.4 (APV84 VLE-IG) and predicted the miscibility gap well. The predicted diffusion coefficients using PC-SAFT and COSMO-RS are rejected because the thermodynamic factor is not positive definite resulting in not positive definite diffusion coefficients, *i.e.*, the predicted miscibility gap is to large.

The results are shown in Fig. 4.31 for the Wesselingh and Krishna (WK) extension to the Vignes equation (A.3) and for the Darken-KvB (A.14) using the thermodynamic factor from NRTL-parameters. Using any of the three models to calculate the thermodynamic factor, WK leads to the lowest error of prediction and the Darken-

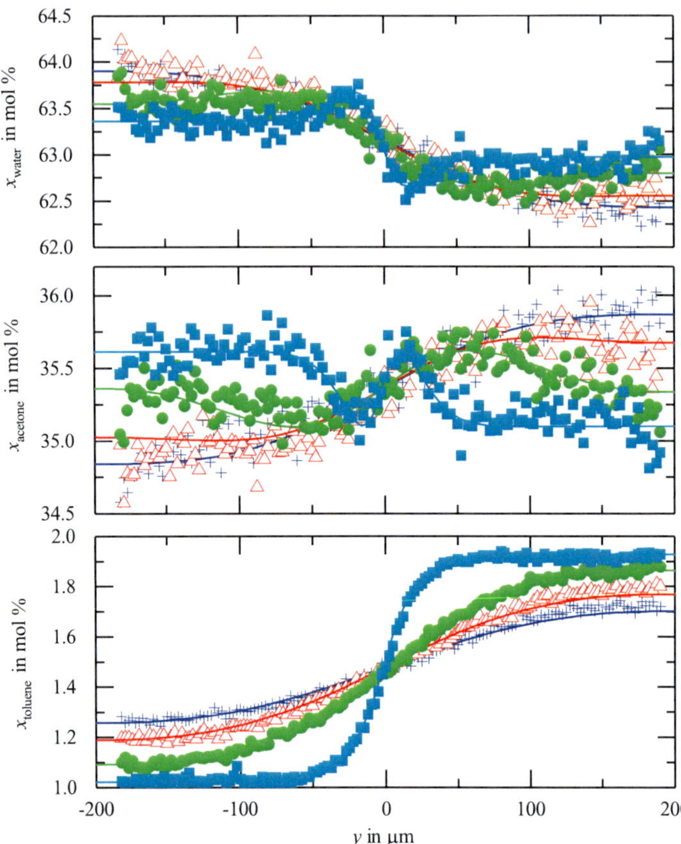

Figure 4.29: Measured (symbols) and fitted (lines) mole fraction profiles of the system water (1) + acetone (2) + toluene along measurement line y of the channel cross-sections at observation points s_1 (■,–), s_2 (•,–), s_3 (△,–), and s_4 (+,–).

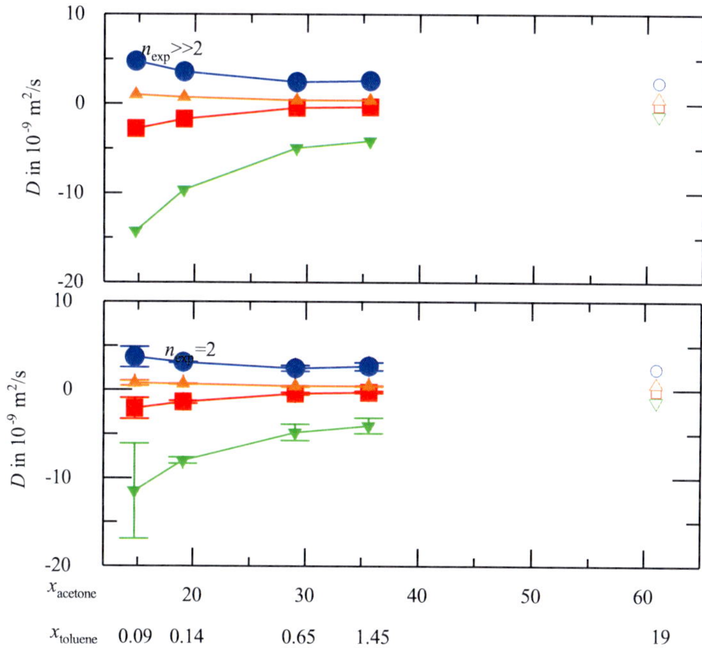

Figure 4.30: Diffusion coefficient matrix $\boldsymbol{D}^V$ with standard deviations of water (1) + acetone (2) + toluene at $T = 25\,°\mathrm{C}$ from this study (filled symbols) and from Krishna *et al.*[54] (empty symbols). The diffusion coefficient matrix consists of D_{11} (■), D_{12} (▼), D_{21} (▲), and D_{22} (●). (Top) The diffusion coefficient is fitted simultaneously to all experiments. (Bottom) The diffusion coefficient is evaluated using $n_{exp} = 2$, with standard deviations from repeated evaluations.

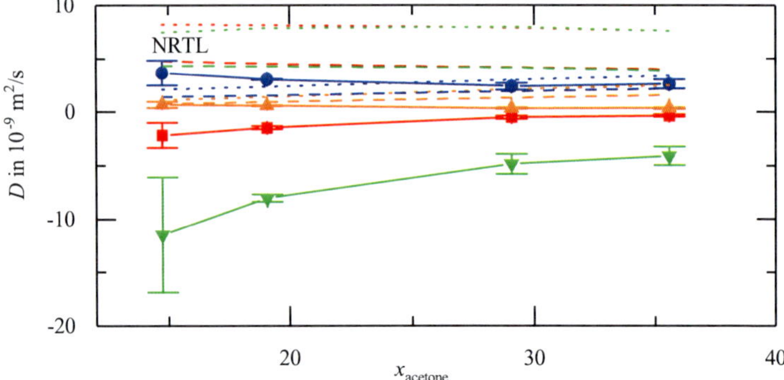

Figure 4.31: Prediction of the diffusion coefficient matrix of water (1) + acetone (2) + toluene at 25 °C using the WK extension to the Vignes equation (- -) and the Darken-KvB (···). The diffusion coefficient matrix consists of D_{11} (■), D_{12} (▼), D_{21} (▲), and D_{22} (•). The thermodynamic factor was calculated from COSMO-RS (top), PC-SAFT (middle), and NRTL (bottom).

KvB to the highest error of all predictive multicomponent models presented in Appendix A.1. The WK extension showed the best results, although it only considers interactions between two components. While WK is significantly different, all other models predict results similar to the Vignes-WK.

Calculating the thermodynamic factor using COSMO-RS resulted in the lowest error ($\sim 2.54 \times 10^{-9}\,\mathrm{m^2\,s^{-1}}$), though a systematic error due the not positive definite thermodynamic factor. All signs in the predicted diffusion coefficient agree with the measured diffusion coefficient. Good agreement is obtained for D_{22}. D_{11} is underestimated, while D_{12}, and D_{21} are overestimated.

The prediction using PC-SAFT showed an average error of $4.87 \times 10^{-9}\,\mathrm{m^2\,s^{-1}}$ for the WK extension of the Vignes-equation and an average error of $6.88 \times 10^{-9}\,\mathrm{m^2\,s^{-1}}$. The prediction of D_{12} shows the wrong sign but a similar value, resulting in the large error. The other three diffusion coefficients are predicted quite well but overall a systematic error due to the miscibility gap is apparent.

Using NRTL parameters resulted in an average error of $4.23 \times 10^{-9}\,\mathrm{m^2\,s^{-1}}$ for the WK extension of the Vignes-equation and an average error of $6.11 \times 10^{-9}\,\mathrm{m^2\,s^{-1}}$. Here, the wrong signs of D_{11}, and D_{12} lead to the large errors. Additionally, the thermodynamic factor from NRTL fails to predict the increasing determinant of the diffusion coefficient matrix $det(\boldsymbol{D}^V)$ with increasing acetone and toluene concentrations.

Overall, all models predict large negative diffusion coefficients, as the experimental results also show. Only COSMO-RS predicts the monotonically increase of the diffusion coefficient matrix $det(\boldsymbol{D}^V)$ with increasing acetone and toluene concentrations, however, within the miscibility gap. A positive definite thermodynamic factor is only predicted by the thermodynamic factor from NRTL. These results show how much the quality of the predicted diffusion coefficient depends on the model used to determine the thermodynamic factor $\boldsymbol{\Gamma}$.

4.8 Quaternary: Cyclohexane + Toluene + Acetone + Methanol

For the quaternary system, the ternary system cyclohexane + toluene + methanol was complemented by acetone. The diffusion coefficient matrix $\boldsymbol{D}^V$ of cyclohexane (1) + toluene (2) + acetone (3) + methanol was measured at 298.15 K and $\bar{x}_1 = \bar{x}_2 = \bar{x}_3 = 5\,\mathrm{mol}\,\%$. The compositions of the solutions used in the experiments are given in Table K.1.

As the spectra of cyclohexane, toluene, and methanol, the spectrum of acetone shows distinctive bands in the fingerprint region, *e.g.*, at $1708\,\mathrm{cm}^{-1}$ and $787\,\mathrm{cm}^{-1}$, see Fig 4.32. The spectral areas of acetone, cyclohexane, and toluene are significantly larger than the spectral area of methanol (60, 40, and 30 times, respectively). From calibration, a small error of mole fractions $\mathrm{RMSE}_{\mathrm{cal}} < 0.23\,\mathrm{mol}\,\%$ is obtained for the quaternary system. The $\mathrm{RMSE}_{\mathrm{cal}}$ is even smaller than the $\mathrm{RMSE}_{\mathrm{cal}}$ of cyclohexane + toluene + methanol.

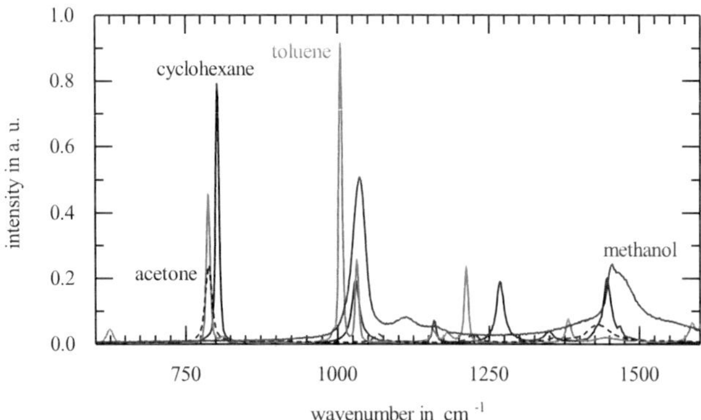

Figure 4.32: Raman spectra of cyclohexane (–), toluene (–), acetone (- -), and methanol (–).

The diffusion coefficient matrix $\boldsymbol{D}^V$ was determined from measured mole fraction profiles. An example of measured and fitted mole fractions from the parameter estimation procedure (3.10) is shown in Fig. 4.33 for one experiment. The agreement between measurements and fits was similar for all experiments. In all experiments, a very good agreement between measured and calculated mole fractions was achieved with $\mathrm{RMSE_{fit}} < 0.06\,\mathrm{mol\,\%}$. The error is similar for the analysis of the diffusion coefficient matrix $\boldsymbol{D}^V$, hardly dependent of the number n_{exp} (Table K.6). The $\mathrm{RMSE_{fit}}$ ranges from $\mathrm{RMSE_{fit}}(n_{\mathrm{exp}}=1)=0.056\,\mathrm{mol\,\%}$ to $\mathrm{RMSE_{fit}}(n_{\mathrm{exp}}=2)=0.051$, $\mathrm{RMSE_{fit}}(n_{\mathrm{exp}}=3)=0.052$, and $\mathrm{RMSE_{fit}}(n_{\mathrm{exp}}=12)=0.047\,\mathrm{mol\,\%}$. Hence, the error of fit $\mathrm{RMSE_{fit}}$ is smaller than the error of calibration $\mathrm{RMSE_{cal}} <0.23\,\mathrm{mol\,\%}$. Even for 16 experiments or 12 experiments with a disadvantageous starting value, a $\mathrm{RMSE_{fit}}$ smaller than $\mathrm{RMSE_{cal}}$ can be obtained (Table K.6). y_0 was smaller than $30\,\mathrm{\mu m}$ in all experiments, and in 13 of 16 experiments y_0 was even smaller than $10\,\mathrm{\mu m}$.

The diffusion coefficient matrix $\boldsymbol{D}^V$ is shown in Fig. 4.34 for $\bar{x}_1 = \bar{x}_2 = \bar{x}_3 = 5\,\mathrm{mol\,\%}$, the numerical data are given in Table K.6. Experiments were evaluated as described above using $n_{\mathrm{exp}}=1$, $n_{\mathrm{exp}}=2$, $n_{\mathrm{exp}}=3$, and $n_{\mathrm{exp}}=12$.

The diffusion coefficient matrix $\boldsymbol{D}^V$ can be estimated from only one single experiment (Table K.5). Generally, one experiment is the minimal experimental effort. As expected, the precision is very low ($u_r(D_{ij}^V) = 113\,\%$ to $541\,\%$), as the entries D_{ij}^V of the diffusion coefficient are correlated ($r = -0.98$ to 0.99). The elements of the diffusion coefficient matrix $\boldsymbol{D}^V$ determined from $n_{\mathrm{exp}}=1$ differ by up to three orders of magnitude between different experiments. Even the determinant of the diffusion coefficient matrix $\boldsymbol{D}^V$ differs by up to two orders of magnitude. The arithmetic mean of all estimated diffusion coefficients $\boldsymbol{D}^V$ is not positive definite, while the individual experiments are. Overall, the estimation of $n_{\mathrm{exp}}=1$ is feasible, but not of practical importance.

The precision of the estimated diffusion coefficient $\boldsymbol{D}^V$ increases rapidly as the number of experiments n_{exp} in the estimation procedure increases. This was already predicted from optimal experimental design techniques for ternary experiments.[32] When the diffusion coefficient $\boldsymbol{D}^V$ is estimated from two independent experiments ($n_{\mathrm{exp}}=2$), the relative standard deviation is reduced from $u_r(\boldsymbol{D}^V) = 116\,\%$ to $u_r(\boldsymbol{D}^V) = 52\,\%$. For $n_{\mathrm{exp}}=3$, the uncertainty $u_r(\boldsymbol{D}^V)$ is approximately $7\,\%$. The maximum observed uncertainty was $15\,\%$ for $u_r(D_{31})$ and $u_r(D_{32})$, each. The averaged diffusion coefficient matrix $\boldsymbol{D}^V$ is positive definite for $n_{\mathrm{exp}}=3$ as well as

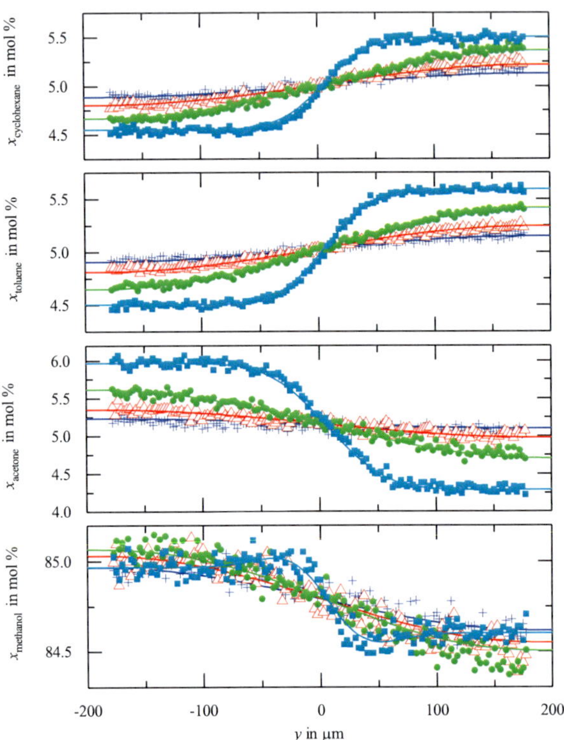

Figure 4.33: Measured (symbols) and fitted (lines) mole fraction profiles of the system cyclohexane (1) + toluene (2) + acetone (3) + methanol for $\bar{x}_1 = \bar{x}_2 = \bar{x}_3 = 5\,\text{mol}\,\%$ along measurement line y of the channel cross-sections at observation points s_1 (■), s_2 (•), s_3(▲), and s_4 (+).

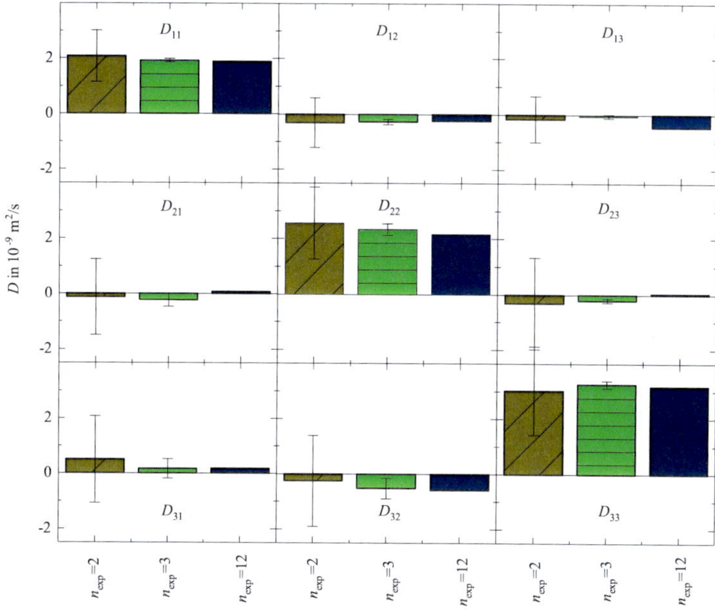

Figure 4.34: Quaternary diffusion coefficient matrix $\boldsymbol{D}^V$ of cyclohexane (1) + toluene (2) + acetone (3) + methanol at $\bar{x}_1 = \bar{x}_2 = \bar{x}_3 = 5\,\mathrm{mol}\,\%$ and $T = 298.15\,\mathrm{K}$. Results are shown for evaluation of n_{exp} experiments with $n_{\mathrm{exp}} = 2$ (■), $n_{\mathrm{exp}} = 3$ (■) and $n_{\mathrm{exp}} = 12$ (■).

$n_{\text{exp}} = 2$.

The diffusion coefficient matrix $\boldsymbol{D}^V$ was also determined from $n_{\text{exp}} = 12$ experiments simultaneously. The results agree within the measurement uncertainty with the diffusion coefficients of $n_{\text{exp}} = 2$. Additionally, the deviations between the results from $n_{\text{exp}} = 12$ and $n_{\text{exp}} = 3$ are small. The determined diffusion coefficient matrix $\boldsymbol{D}^V$ describes the measured mole fraction profiles very well for $n_{\text{exp}} = 12$, *i.e.*, an $\text{RMSE}_{\text{fit}} = 0.047$ is obtained.

The quaternary diffusion coefficient matrix was also predicted using engineering models presented in Appendix A.1 and COSMO-RS as well as PC-SAFT. Using PC-SAFT to determine the thermodynamic factor, the association interaction was considered for the binary subsystem acetone + methanol with $\kappa^{A_iB_j}$=0.035 176 of methanol for both components, and $\epsilon^{A_iB_j} = 0$. Overall, the agreement between the measured and the predicted diffusion coefficient matrix was higher for the thermodynamic factor determined from PC-SAFT and only the thermodynamic factor predicted using PC-SAFT is positive definite.

Therefore, measured and predicted diffusion coefficients using Vignes-WK (A.3), Vignes-KVB (A.5), Darken-KvB (A.14), and Darken-LBV (2.18) in combination with PC-SAFT are shown in Fig. 4.35. The smallest error was obtained for Vignes-WK (33 %) and the Darken-LBV model (34 %). The highest errors were determined for the Vignes-KVB model, 45 % using COSMO-RS and 53 % for PC-SAFT. For this system, the differences between the engineering models are smaller than the differences resulting from PC-SAFT and COSMO-RS. Generally, D_{11} and D_{22} are overestimated while D_{33} is underestimated. The signs of five of six cross-diffusion coefficients are predicted correctly, though the standard deviations of D_{13}, D_{21}, and D_{31} cross 0. The values are similar from measurements and predictions for D_{13}, D_{23}, D_{31}, and D_{32}, while D_{12}, and D_{21} differ significantly regarding the values from measurements and predictions.

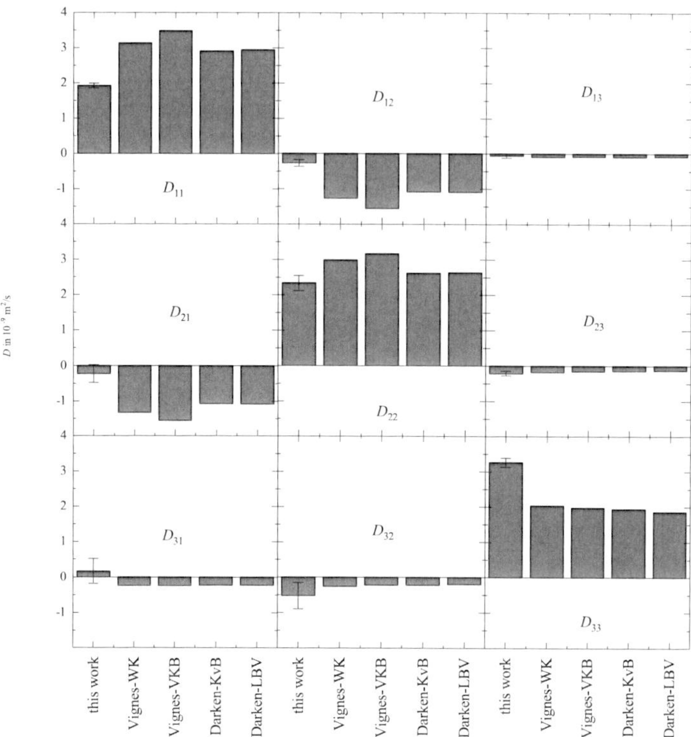

Figure 4.35: Diffusion coefficient matrix $\boldsymbol{D}^V$ of cyclohexane (1) + toluene (2) + acetone (3) + methanol at $T = 25\,^\circ\mathrm{C}$ from this study with $n_{\mathrm{exp}} = 3$ in comparison to predicted diffusion coefficient matrices with the thermodynamic factor $\boldsymbol{\Gamma}$ obtained from PC-SAFT and MS diffusion coefficients obtained from the generalized Vignes equation with WK and VKB extensions as well as the Darken-KvB and the Darken-LBV models.

4.9 Discussion of Measured and Predicted Multicomponent Diffusion Coefficients in Liquids

The results of the previous sections (Sec. 4.1-4.8) show that the prediction of diffusion coefficients is still difficult, for binary, ternary, and quaternary systems. While the thermodynamic factor based on PC-SAFT was well suited for cyclohexane + toluene, and acetone + toluene, COSMO-RS showed superior predictions for most other systems. However, COMO-RS and PC-SAFT predicted the miscibility gap of water + acetone + toluene too large. AS the PC-SAFT parameters were not fitted to phase equilibria, the results might improve with $k_{ij} \neq 0$.

For binary systems, the Vignes-eq. showed promising results. For the ternary and quaternary systems measured in this work, no distinct recommendation regarding the multicomponent engineering models is possible. Darken-LBV and Darken-KvB both show promising results, while also the multicomponent Vignes extensions provide very similar results. Surprisingly, the Vignes-WK leads to comparable results for cyclohexane + toluene + methanol and the quaternary system, while it has the lowest prediction accuracy for 1-propanol + 1-chlorobutane + heptane. Therefore, reliable experimental methods with short experiment time are still required for the determination of diffusion coefficients.

For all systems, a good agreement between the convection-diffusion model (3.1)-(3.4) and the measurements was achieved. Therefore, the assumptions in the convection-diffusion model are sensible and the model suitable for the procedure. Neither was a significant shift of the interface ($> 30\,\mu m$) due to viscosity observed, nor increased diffusivities due to a rotation of the interface. Also, mixing effects as excess volumes or locally different diffusivities due to temperature fluctuations can be excluded. As all solutions are transparent in visible light, heating due to absorption of the laser is negligible. An influence of the spiral geometry of the microfluidic chip was not observed, neither in this work nor by Haase[235] or Juring[236]. As $De^2 Sc \ll 100$, the spiral geometry was *a priori* unlikely to influence the diffusion experiments.

Sources for statistical errors in the diffusion experiments in this work are the volume flow, the sample preparation and the temperature in the microchannel. Though the volume flow would be a major error source, the influence is low as the uncertainty of the volume flow is low (0.25 %). This uncertainty of the volume flow results in a statistical error of 0.17 %.[235] Despite the step motor of the syringe pump, the assumption of steady state is plausible due to inertial and capillary forces.

The sample preparation is another error source in the experimental procedure: Though the balance used for sample preparation has an accuracy of ±1 mg, additional errors are introduced by the high vapor pressures of all solvents within this work. To reduce these errors, code of best practice for sample preparation was complied with. Additionally, statistical errors within this work were compensated by the evaluation procedure. The inlet concentrations of the diffusion experiments were also estimated in the fitting procedure and conflicting inlet concentrations were rejected in the concentration calibration procedure. As the $\mathrm{RMSE}_{\mathrm{fit}}$ was similar or smaller than the $\mathrm{RMSE}_{\mathrm{cal}}$, this approach is reasonable.

Possible concentration offsets in the sample preparation can result in errors for systems with a large dependence of the diffusion coefficient on the concentration. A deviation of ±0.1 mol % in the overall concentration of, *e.g.*, cyclohexane + toluene would result in an error of 0.7 %.

The temperature uncertainty in this work is ±0.5 K due to the temperature control procedure. Thus, the uncertainty of the diffusion coefficient due to the temperature uncertainty was halved compared to the work of Haase.[235]

Uncertainties in sample preparation, volume flow and temperature result in a statistical error of 2.56 % for the diffusion coefficients of cyclohexane + toluene measured in this work. This agrees well with a relative error of 2.4 % determined from repeated experiments for cyclohexane + toluene. Additionally, Haase has determined a similar maximal statistical measurement error for cyclohexane + toluene (3 %).

Systematic errors result from small deviations in the geometry of the microfluidic chip from the original chip design (±0.3 µm) in this work. The width of the microfluidic chip changes the volume flow linearly but also the diffusion width within the convection-diffusion model. A difference in the height also changes the volume flow, but has no further influence in the convection-diffusion model. Due to the large area of width to height, an error of 1 µm in the height has the tenfold influence on volume flow compared to the width. These two errors could add up to 3 %.[235]

The distances from the observation points s_1 to s_4 to the position where the two channel inlets join have been determined in the technical drawing. Here, errors up to 0.1 mm appear plausible and would result in an additional error of 3 %.[235] This error was reduced to approximately 1.5 % with the additional fitting parameter Δs. If the distance between the observation points was incorrect, Δs would still be used to find the lowest $\mathrm{RMSE}_{\mathrm{fit}}$ and not the optimal first concentration profile.

Thus, a systematic error of up to 4.5 % is plausible. However, the diffusion coefficients measured in this work agree well with literature data (apart from acetone + water at x_{acetone} =59.95 mol %). Therefore, either the errors in the measured geometry counterbalance each other or the real errors are smaller than estimated errors.

Deviations between measurements in this work and literature data can also be caused by different grades of the used chemicals. HPLC grade for example does not define the quality unambiguously and Tominaga *et al.*[208] have shown that diffusion coefficients can differ significantly for different alcohols, *e.g.*, impurities in the used solvents. However, these systematic deviations cannot be estimated.

Overall, Raman microspectroscopy in an H-cell microchannel allows the measurement of binary, ternary, and quaternary diffusion coefficients with low effort, low uncertainties and high precision. To ensure reliable diffusion coefficients, it is important that the experiments comply to the mentioned dimensionless quantities and characteristics (Sec. 3.4). Changes of the geometry of the microfluidic chip should be considered to avoid violations of these quantities and characteristics. Additionally, the chemical system needs to be Raman active and the uncertainty of mole fraction measurements should be smaller than the concentration difference between the inlet solutions. In conclusion, no general limitations in the system were observed, which would deteriorate the measurement of diffusion coefficients in systems with more than four components.

Chapter 5

Summary, Conclusions and Future Perspective

A measurement method was developed in this work for multicomponent diffusion coefficients in liquids by combining microfluidics with Raman microspectroscopy. While the feasibility of binary diffusion measurements using microfluidics and Raman microspectroscopy has already been shown and the potential for multicomponent diffusion measurements was hypothesized,[5,31,36,76] the feasibility of microfluidic multicomponent diffusion measurements was only shown in this work and associated publications.[141,250–252]

The experiments were validated using the binary systems cyclohexane + toluene, acetone + toluene, and acetone + water, and the ternary systems 1-propanol + 1-chlorobutane + heptane, and cyclohexane + toluene + methanol. New diffusion data was measured for the systems cyclohexane + methanol, water + acetone + toluene and cyclohexane + toluene + acetone + methanol. For all systems, all experiments were well described by the multicomponent convection-diffusion model. The deviations between model and measurement are smaller or in the same order of magnitude as the error of the mole fraction measurements. Therefore, the assumptions underlying the convection-diffusion model created in this thesis agree with the experimental results.

The measured diffusion coefficients agree with the literature values within the measurement uncertainty, except for acetone + toluene and D_{12} and D_{22} of the system cyclohexane + toluene + methanol. This agreement between results from this work and literature data also supports the validity of the convection-diffusion model for the described diffusion experiments. Overall, reliable operation has been proven for binary, ternary, and quaternary systems. Here, the ternary and quaternary systems were used as prototype of multicomponent systems. As the measurement method of Raman spectroscopy provides information on mole fractions of all components simultaneously, the measurement method is expected to provide also diffusion coefficients for systems with more components.

It was shown that one multicomponent experiment suffices to determine ternary and quaternary diffusion coefficients. For the ternary system 1-propanol + 1-chlorobutane + heptane, a reasonable accuracy of the diffusion coefficients ($\sigma(\boldsymbol{D}^V) = 33\,\%$) was already achieved from a single experiment. By combining two independent experiments, excellent precision and accuracy were achieved ($\sigma(\boldsymbol{D}^V) = 3.1\,\%$). Similar observations were also made for cyclohexane + toluene + methanol and water + acetone + toluene. For the quaternary system, three independent diffusion experiments should be performed, to obtain an accuracy of 6 % or higher. Thus, at least $n_c - 1$ experiments should be performed in a mixture containing n_c components.

Despite the unnecessarily high spatial resolution of the Raman spectra (2.5 µm), the experiment duration (< 1 h) is competitive to the fastest classical diffusion experiments. The sample consumption is less than 5 mL. Furthermore, the number of experiments was reduced in comparison to classical methods.

The accuracy of the diffusion measurement as well as the error of the mole fraction calibration were specific for every system and every concentration. Overall, a high accuracy of the measured diffusion coefficient was observed, *e.g.*, for cyclohexane + toluene ($u_r = 2.4\,\%$). The influence of small temperature variations (± 0.5 K) is expected to have an influence of $\pm 1.7\,\%$ on the diffusion coefficient. For smaller temperature fluctuations and higher accuracy of mole fraction calibrations, even a higher accuracy of the diffusion coefficient can be expected. Therefore, a recording of the Pt100 and the thermocouple temperature data is interesting together with further measures to narrow the temperature uncertainty.

The current model assumed negligible excess volumes and negligible change in viscosities between the inlet solutions. These limitations could probably be overcome by expanding the convection-diffusion model, regarding local viscosities and significant changes of the partial molar volumes.[166]

The main limitation of the presented setup is that the studied species must be Raman active and sufficiently different in the Raman spectra. If this is obeyed, even diffusion measurements at infinite dilution are possible, *e.g.*, cyclohexane in methanol. The analysis is difficult, when the accuracy of the mole fraction measurements is similar to the mole fraction difference between the inlet solutions or less. Interferences as fluorescence could be handled, *e.g.*, by using a different laser wavelength to suppress fluorescence or pseudo-second derivatives in the analysis to make the evaluation more robust regarding background signals.[229,253–255] If the Raman signal of one component cannot be separated from the background signal using IHM, *e.g.*, low water contents in organic solutions, other spectral analysis techniques might be of interest, *e.g.*, very hard modeling[256] or partial least squares (PLS).[257] If Raman spectroscopy is not suitable for a mixture of interest, other spectroscopic techniques such as Fourier transform infrared spectroscopy[6,258] could be used in the setup as long as they allow for *in situ* confocal measurements.

The setup could also be used for diffusion coefficient measurements in more complex fluids, *e.g.*, large molecules or particles in solution or viscous solvents. These measurements should be possible with adaptions of the microchannel geometry, *e.g.*, to account for the very small diffusion coefficients in the order of $10^{-10}\,\mathrm{m^2\,s^{-1}}$ to $10^{-11}\,\mathrm{m^2\,s^{-1}}$.

The efficiency of this method could be further improved by a model-based experimental procedure: As could be seen for cyclohexane + toluene, acetone + toluene and acetone + water, the acquisition of Raman spectra every 2.5 µm provides much more information than necessary. Here, a tradeoff is to be exploited between accuracy of the diffusion coefficient and the number of measurements.

As shown for 1D Raman diffusion measurements, the information content is higher near the walls. On the other hand, the measurement accuracy of the mole fractions is low near the walls. Model-based experimental procedure could identify the most interesting y-positions as well as non-uniformly spaced y-grids for the Raman spectra acquisition.

It was also shown for cyclohexane + toluene that only three observation points provide enough information to determine the diffusion coefficient with high accuracy. The least uncertainty was found for the combination of s_1, s_2, and s_4. As Fo_{opt} was determined for the inline measurement of diffusion coefficients, Fo_{opt} for *in situ* measurement could differ. Thus, it should be addressed whether there are superior observation points s, which are accessible in the geometry of this setup.

The output of the measurement method could be increased by automating the setup with independent, programmable syringe pumps which premix the inlet solutions in a microfluidic mixer. This premixing with independently tunable volume flows allows repeated experiments and to steer diffusion experiments subsequently through a wide concentration range without lag times for the effusion of air bubbles. The feasibility of this premixing has already been tested in a student research project by Münch *et al.*[259]

Without any experimental adaptions, the setup would also allow for the characterization of reaction-diffusion systems.[12,133,260,261] By using immiscible fluids and increasing the retention time, even the determination of liquid-liquid equilibria is possible.[262] By using more complex microfluidic chips with intersecting microchannels, parallel operation of diffusion measurements on one chip could be possible.[12,263,264] These microfluidic chips could provide several parallel or subsequent microchannels to observe diffusion while upstream microchannels could premix the inlet solutions for the diffusion measurements.

More important, however, is to make microfluidic multicomponent diffusion measurements accessible. For this purpose, the microfluidic setup could be combined with offline sample analysis instead of Raman microspectroscopy. This was successfully operated by Häusler *et al.* for binary diffusion coefficients,[36] and adapted in a simple manner to multicomponent mixtures within previous work by Peters *et al.*[141,250]

Overall, the presented combination of microfluidics and Raman microspectroscopy is a rapid and reliable method for diffusion measurements in multicomponent mixtures.

Appendix A

Prediction of Binary and Multicomponent Diffusion Coefficients

The prediction of multicomponent diffusion coefficients is introduced shortly in Sec. 2.3 for concentrated solutions. In this Appendix, further engineering models for binary and multicomponent mixtures are presented (Sec. A.1) as well as models for the temperature-dependence of diffusion coefficients (Sec. A.2)

A.1 Concentration-Dependence of Diffusion Coefficients

Predictive models for Maxwell-Stefan diffusion coefficients in concentrated (multicomponent) solutions include the Vignes equation for binary mixtures[265],

$$Đ_{\mathrm{AB}} = (Đ_{\mathrm{AB}}^{x_{\mathrm{A}}\to 0})^{x_{\mathrm{B}}} (Đ_{\mathrm{AB}}^{x_{\mathrm{B}}\to 0})^{x_{\mathrm{A}}} , \tag{A.1}$$

and the generalized (multicomponent) Vignes equation for multicomponent mixtures[266]

$$Đ_{ij} = (Đ_{ij}^{x_i\to 1})^{x_i} (Đ_{ij}^{x_j\to 1})^{x_j} \prod_{k=1,k\neq i,j}^{n} (Đ_{ij}^{x_k\to 1})^{x_k} . \tag{A.2}$$

Various approaches based on binary diffusivities have been suggested to estimate $Đ_{ij}^{x_k \to 1}$, *e.g.*, Wesselingh and Krishna (WK)[266]

$$Đ_{ij}^{x_k \to 1} = \sqrt{Đ_{ij}^{x_j \to 1} Đ_{ij}^{x_i \to 1}} \,, \tag{A.3}$$

Kooijman and Taylor (KT)[59]

$$Đ_{ij}^{x_k \to 1} = \sqrt{Đ_{ik}^{x_k \to 1} Đ_{jk}^{x_k \to 1}} \,, \tag{A.4}$$

Krishna and van Baten (VKB)[267]

$$Đ_{ij}^{x_k \to 1} = (Đ_{ik}^{x_k \to 1})^{x_i/(x_i+x_j)} (Đ_{jk}^{x_k \to 1})^{x_j/(x_i+x_j)} \,, \tag{A.5}$$

Krishna and van Baten (DKB)[267]

$$Đ_{ij}^{x_k \to 1} = \frac{x_j}{x_i + x_j} Đ_{ik}^{x_k \to 1} + \frac{x_i}{x_i + x_j} Đ_{jk}^{x_k \to 1} \,, \tag{A.6}$$

Rehfeldt and Stichlmair (RS)[241,268]

$$Đ_{ij}^{x_k \to 1} = (Đ_{ik}^{x_k \to 1} Đ_{jk}^{x_k \to 1} Đ_{ij}^{x_j \to 1} Đ_{ij}^{x_i \to 1})^{1/4} \,, \tag{A.7}$$

and Liu, Bardow, and Vlugt (Vignes-LBV)[26]

$$Đ_{ij} = (D_{j,\text{self}}^{x_i \to 1})^{x_i} (D_{i,\text{self}}^{x_j \to 1})^{x_j} \prod_{k=1, k \neq i,j}^{n} \left(\frac{D_{i,\text{self}}^{x_k \to 1} D_{j,\text{self}}^{x_k \to 1}}{D_{k,\text{self}}^{x_k \to 1}} \right)^{x_k} . \tag{A.8}$$

Darken presented another model for binary systems[68]

$$Đ_{ij} = x_i D_{j,\text{self}}(x_i) + x_j D_{i,\text{self}}(x_i) \tag{A.9}$$

already in 1948. This model was extended by Liu, Bardow, and Vlugt to the multi-component Darken-LBV equation[26]

$$Đ_{ij} = \frac{D_{i,\text{self}} D_{j,\text{self}}}{D_{\text{mix}}} \tag{A.10}$$

with

$$D_{\text{mix}} = \sum_{i=1}^{n} \frac{x_i}{D_{i,\text{self}}} \tag{A.11}$$

and the predictive Darken-LBV[26] extension using

$$\frac{1}{D_{i,\text{self}}} = \sum_{j=1}^{n} \frac{x_j}{D_{i,\text{self}}^{x_j \to 1}} \ . \tag{A.12}$$

In infinite dilution, the diffusion coefficients

$$D_{i,\text{self}}^{x_j \to 1} = D_{ij}^{x_j \to 1} = Đ_{ij}^{x_j \to 1} \tag{A.13}$$

are equal.

A competing approach is the Darken-KvB[267]

$$Đ_{ij} = \frac{x_i}{x_i + x_j} D_{j,\text{self}} + \frac{x_j}{x_i + x_j} D_{i,\text{self}} \tag{A.14}$$

with

$$D_{i,\text{self}} = \sum_{j=1}^{n} w_j D_{i,\text{self}}^{w_j \to 1} \ . \tag{A.15}$$

A.2 Temperature-Dependence of Diffusion Coefficients

Several models for the temperature-dependence of diffusion coefficients were compiled by Poling *et al.*[50] These include models based on the Wilke-Chang eq. (2.16) for infinite dilution, recommended for small temperature ranges

$$\frac{D_{\mathrm{AB}}^{x_{\mathrm{A}}\to 0}\eta_{\mathrm{B}}}{T} = \text{constant} . \tag{A.16}$$

Another model to predict the temperature-dependence at infinite dilution is, *e.g.*, Tyn[269]

$$\frac{D_{\mathrm{AB}}^{x_{\mathrm{A}}\to 0}(T_2)}{D_{\mathrm{AB}}^{x_{\mathrm{A}}\to 0}(T_1)} = \left(\frac{T_{\mathrm{c,B}} - T_1}{T_{\mathrm{c,B}} - T_2}\right)^n , \tag{A.17}$$

with n depending on the heat of vaporization.[50] The temperature-dependence for concentrated solutions and infinite dilution can be modeled like the Arrhenius eq. using[270]

$$D_{\mathrm{AB}} = A \exp \frac{-B}{T} . \tag{A.18}$$

To predict the temperature-dependence of Fick diffusion coefficients in concentrated solutions, also models based on the Wilke-Chang eq. are possible[70]

$$D_{AB}(x, T) = D_{AB}(x, T_0) \frac{T \eta_{AB}(x, T_0)}{T_0 \eta_{AB}(x, T)} . \tag{A.19}$$

Appendix B

Diffusion-Convection Models

As discussed in Sec. 2.6.4, diffusion-convection models are available to analyze microfluidic diffusion measurements starting from 3D models combined with FEM to analytical solutions for 1D problems. In the following, examples for the different diffusion-convection models are introduced.

3D diffusion-convection models are widespread for the evaluation of diffusion experiments. The 3D models start with a combination of the Navier-Stokes equation for pressure-driven flow with 3D diffusion, where velocity and concentration profiles are simultaneously solved using FEM.[18,165,166] Yang *et al.*[8] solved a simplified model with 3D flow and 2D diffusion in s- and y-direction using CFD (for $H/W < 0.5$). A similar problem was solved by Song *et al.*[150] using a Fourier series solution for 2D diffusion in y- and z-direction with a 3D Navier-Stokes equation neglecting gravitational effects. Loureiro *et al.*[169] solved a 3D diffusion-convection model using FEM for the Taylor's pulse dispersion in a different microfluidic experimental setup.

The second group of diffusion-convection models is reduced to 2D. Abonnenc *et al.*[133] evaluated a convection-diffusion-reaction system with $W/H > 10$ with a 2D model considering convection in s-direction and diffusion in y-direction using FEM. Another 2D diffusion-convection model was used to describe diffusive peak broadening.[120] Wu *et al.*[127,143] applied a 2D analytical model for diffusion in y- and s-direction using $v(y) = \text{const.}$ to determine diffusion coefficients and to evaluate differences in the analysis using either concentration-dependent diffusion coefficients or the assumption $D = \text{const.}$ Other 2D models were solved via Fourier analysis[170] or using a series of eigenfunctions.[159] Free diffusion in y-direction was solved with a $v(s, y)$-velocity profile due to viscosity changes using analytical and numerical solutions.[149] Kamholz and Yager[134] solved a 2D diffusion-convection model with a parabolic velocity profile numerically for restricted diffusion y-wise and free diffusion z-wise. Häusler *et al.*[36] measured diffusion coefficients in a restricted y-diffusion geometry, see also eqn (2.41)-(2.43).

1D convection-diffusion models are generally simplified to a constant velocity v. The constant velocity is often matched by diffusion measurements in a free diffusion constellation.[38,76,168] Jeon *et al.*[7] and Holden *et al.*[167] described special cases of 1D models with analytical solutions. Hotta *et al.*[132] applied a penetration model for free diffusion to analyze diffusion experiments. Other experiments required either a model for reaction-diffusion-convection systems using pseudo-binary free diffusion,[138] or a model for magnetophoresis-diffusion in T-sensor,[119] or a model for restricted diffusion across liquid-liquid interfaces.[271]

Overall, for modeling diffusion in microfluidics and evaluation of microfluidic diffusion measurements, there is not one solution that fits all experimental approaches. 3D FEM models are generally feasible, but are complex, require computational and preparation time and risk numerical errors. 1D models and analytical solutions reduce complexity, computational time and requirements, but are not available for all experimental approaches or risk significant errors. The model presented by Häusler *et al.* and in this work proved suitable for the described task of evaluating diffusion measurements with the applied mesh sizes and optimization procedure including termination criteria. The computational times were a few seconds for the evaluation of one binary diffusion experiment to several hours for twelve quaternary diffusion experiments.

Appendix C

Parametrization of the Quaternary Diffusion Coefficient Matrix in the Estimation Procedure

To avoid violations of these constraints during the estimation of the diffusion coefficient and to ensure a positive definite diffusion coefficient, the following parametrization $\boldsymbol{\theta}$ is used for quaternary diffusion coefficients to be able to use simple constraints on the diffusion coefficient.
The eigenvalues of the diffusion coefficient matrix are estimated

$$\theta_1 = D_1^\dagger\,, \quad \theta_2 = D_2^\dagger\,, \quad \text{and} \quad \theta_3 = D_3^\dagger \tag{C.1}$$

together with the main diffusion coefficients

$$\theta_4 = D_{11} - D_{22} \quad \text{and} \quad \theta_5 = D_{11} - D_{33} \tag{C.2}$$

and the cross-diffusion coefficients

$$\theta_6 = D_{11} - D_{21}\,, \quad \theta_7 = D_{11} - D_{23}\,, \quad \theta_8 = D_{11} - D_{31}\,, \quad \text{and} \quad \theta_9 = D_{11} - D_{32}\,. \tag{C.3}$$

The parametrization is transformed back to the main diffusion coefficients by

$$D_{11} = \frac{-(\theta_2-\theta_3)(\theta_1-\theta_3)(\theta_1-\theta_2)(\theta_4+\theta_5+\theta_1+\theta_2+\theta_3)}{(\theta_2-\theta_3)\theta_1(-\theta_4-\theta_5+\theta_7+\theta_9-2\theta_1)+(-\theta_1+\theta_3)\theta_2(-\theta_4-\theta_5+\theta_7+\theta_9-2\theta_2)+(\theta_1-\theta_2)(\theta_3(-\theta_4-\theta_5+\theta_7+\theta_9-2\theta_3)-(\theta_2-\theta_3)(\theta_1-\theta_3))}, \tag{C.4}$$

$$D_{22} = D_{11} - \theta_4\,, \quad \text{and} \quad D_{33} = D_{11} - \theta_5 \tag{C.5}$$

and to the cross-diffusion coefficients by

$$D_{21} = D_{11} - \theta_6\,, \quad D_{23} = D_{11} - \theta_7\,, \quad D_{31} = D_{11} - \theta_8\,, \quad D_{32} = D_{11} - \theta_9\,, \tag{C.6}$$

$$\begin{aligned} D_{12} = {} & \frac{((D_{22}-\theta_2)D_{31}-D_{21}D_{32})D_{11}(\theta_1^2+(-D_{22}-D_{33})\theta_1+D_{22}D_{33}-D_{23}D_{32})}{(-D_{23}D_{31}^2-D_{21}(D_{22}-D_{33})D_{31}+D^2D_{32})(\theta_1-\theta_2)} \\ & + \frac{((-D_{22}+\theta_1)D_{31}+D_{21}D_{32})D_{11}(\theta_2^2+(-D_{22}-D_{33})\theta_2+D_{22}D_{33}-D_{23}D_{32})}{(-D_{23}D_{31}^2-D(D_{22}-D_{33})D_{31}+D^2D_{32})(\theta_1-\theta_2)} \\ & + \frac{((-D_{22}+\theta_2)D_{31}+D_{21}D_{32})\theta_1(\theta_1^2+(-D_{22}-D_{33})\theta_1+D_{22}D_{33}-D_{23}D_{32})}{(-D_{23}D_{31}^2-D_{21}(D_{22}-D_{33})D_{31}+D^2D_{32})(\theta_1-\theta_2)} \\ & - \frac{((-D_{22}+\theta_1)D_{31}+D_{21}D_{32})\theta_2(\theta_2^2+(-D_{22}-D_{33})\theta_2+D_{22}D_{33}-D_{23}D_{32})}{(-D_{23}D_{31}^2-D_{21}(D_{22}-D_{33})D_{31}+D_{21}^2D_{32})(\theta_1-\theta_2)} \end{aligned} \tag{C.7}$$

and

$$D_{13} = \frac{D_{11}((D_{22}-\theta_1)(D_{33}-\theta_1)-D_{23}D_{32})-\theta_1((D_{22}-\theta_1)(D_{33}-\theta_1)-D_{23}D_{32})+D_{12}(D_{23}D_{31}-D_{21}(D_{33}-\theta_1))}{-D_{21}D_{32}+(D_{22}-\theta_1)D_{31}} \tag{C.8}$$

Appendix D

Results: Cyclohexane + Toluene

In the following, the concentrations of the inlet solutions of cyclohexane + toluene (Table D.1) are presented together with the calculation of the partial molar volumes (Table D.2), the determined diffusion coefficients (4.1) and the predicted uncertainty due to temperature uncertainties in the diffusion experiments (Table D.4).

Table D.1: Composition of the experiments with cyclohexane + toluene in mol %

	$\bar{x}$	x_l	x_r
cyclohexane	60	58.5	61.5
toluene	40	41.5	38.5

The excess volume of cyclohexane (1) + toluene is described by the Redlich-Kister type function

$$V_E/(\mathrm{cm^3\,mol^{-1}}) = x_1 x_2 (2.294 + 0.496(x_1 - x_2)) \ , \tag{D.1}$$

based on data in ref.[201] This Redlich-Kister type function results in the partial molar volumes given in Table D.2.

Diffusion coefficients for all four diffusion experiments of cyclohexane + toluene are given in Table D.3. The table includes evaluations with $n_{\mathrm{exp}} = 1$ as well as $n_{\mathrm{exp}} = 4$.

The uncertainty of the diffusion coefficient regarding the temperature uncertainty is predicted using the temperature-dependence correlation in concentrated solutions

Table D.2: Partial molar volumes of cyclohexane + toluene

$x_{\text{cyclohexane}}$ in mol %	$\bar{V}_{\text{cyclohexane}}$ in $10^{-6}\,\text{m}^3\,\text{mol}^{-1}$	$\bar{V}_{\text{toluene}}$
60	109.3	107.6

Table D.3: Measured diffusion coefficient for the binary system cyclohexane + toluene at $x_{\text{cyclohexane}} = 60\,\text{mol}\,\%$ and $25\,°\text{C}$ from microfluidic experiments in this work with standard deviation $\sigma(\boldsymbol{D}^V)$. The quality of fit for the diffusion coefficient is given as RMSE_{fit}.

	$\boldsymbol{D}^V$ in $10^{-9}\,\text{m}^2\,\text{s}^{-1}$	RMSE_{fit} in mol %
V1	1.816	0.088
V2	1.730	0.079
V3	1.763	0.087
V4	1.737	0.083
$n_{\text{exp}} = 1$	1.762(39)	0.084
$n_{\text{exp}} = 4$	1.764	0.084
Sanni *et al.*[172]	1.767(35)	

(A.19). For a temperature accuracy of ± 0.5 K, a relative error of the diffusion coefficient of 0.8 % is expected according to the temperature correlation (A.19) and the viscosities of cyclohexane + toluene listed in Table D.4. Thus, the influence of the temperature inaccuracy on the diffusion coefficient is expected to be in the same order of magnitude as the relative error $u_r(D) = 2.2\,\%$. Other uncertainties result from the sample preparation as well as the volume flow.

Table D.4: Expected influence of temperature on the diffusion coefficient and the viscosity of cyclohexane + toluene at $x_{\text{cyclohexane}} = 60\,\text{mol}\,\%$. Viscosities were interpolated based on the data published by Silva *et al.*[272] The diffusion coefficient at 25 °C was published by Sanni *et al.*[172]

T in °C	η in mPa s	D in $10^{-9}\,\text{m}^2\,\text{s}^{-1}$	comment on D
24.5	0.635	1.753	eq. A.19
25	0.631	1.767	Reference[172]
25.5	0.627	1.781	eq. A.19

Appendix E

Results: Acetone + Toluene

In the following, the concentrations of the inlet solutions of acetone + toluene (Table E.1) are presented together with the calculations of the partial molar volumes (Table E.2), and the determined diffusion coefficients (Table E.3).

Table E.1: Composition of the experiments with acetone + toluene in mol%

	$\bar{x}$	x_l	x_r
acetone	40.5	38.5	42.5
toluene	59.5	61.5	57.5
acetone	59.5	57.5	61.5
toluene	40.5	42.5	38.5

The excess volume of acetone (1) + toluene is described by the Redlich-Kister type function

$$V_E/(\mathrm{cm^3\,mol^{-1}}) = x_1 x_2(-0.648 + 0.0436(x_1 - x_2)) \tag{E.1}$$

based on data in ref.[209] This Redlich-Kister type function results in the partial molar volumes given in Table E.2.

Diffusion coefficients for all diffusion experiments of acetone + toluene are given in Table E.3. The table includes evaluations with $n_{\mathrm{exp}} = 1$ as well as $n_{\mathrm{exp}} = 3$.

Table E.2: Partial molar volumes of acetone + toluene

$x_{acetone}$ in mol %	$\bar{V}_{acetone}$ in $10^{-6}\,m^3\,mol^{-1}$	$\bar{V}_{toluene}$
40.5	73.8	106.8
59.5	74.0	106.7

Table E.3: Measured diffusion coefficient for the binary system acetone + toluene at $x_{acetone}$ = 40.5 mol % and $x_{acetone}$ = 59.5 mol % at 25 °C from microfluidic experiments in this work with standard deviation $\sigma(\boldsymbol{D}^V)$. The quality of fit for the diffusion coefficient is given as $\mathrm{RMSE_{fit}}$.

x_1 in mol %		$\boldsymbol{D}^V$ in $10^{-9}\,m^2\,s^{-1}$	$\mathrm{RMSE_{fit}}$ in mol %
	V1	2.604	0.144
	V2	2.421	0.240
40,5	V3	2.480	0.231
	$n_{exp} = 1$	2.502(93)	0.232
	$n_{exp} = 3$	2.499	0.210
	Baldauf and Knapp[212]	2.48(9)	
	V4	2.762	0.156
	V5	2.945	0.144
59,5	V6	2.978	0.163
	$n_{exp} = 1$	2.895(12)	
	$n_{exp} = 3$	2.862	0.156
	Baldauf and Knapp[212]	2.38(8)	

Appendix F

Results: Acetone + Water

In the following, the concentrations of the inlet solutions of acetone + water (Table F.1) are presented together with the calculation of the partial molar volumes (Table F.2), and the determined diffusion coefficients (Table F.3).

Table F.1: Composition of the experiments with acetone + water in mol%

	$\bar{x}$	x_l	x_r
acetone	19.1	18.39	19.81
water	80.9	81.61	80.19

The excess volume of acetone (1) + water is described by the Redlich-Kister type function

$$V_E/(\text{cm}^3\,\text{mol}^{-1}) = x_1x_2(-5.76014 + 2.219(x_1 - x_2) + -0.734(x_1 - x_2)^2) \qquad \text{(F.1)}$$

based on data in ref.[215] This Redlich-Kister type function results in the partial molar volumes given in Table F.2.

Diffusion coefficients for all diffusion experiments of acetone + water are given in Table F.3. The table includes evaluations with $n_{\text{exp}} = 1$ as well as $n_{\text{exp}} = 3$.

Table F.2: Partial molar volumes of acetone + water

$x_{acetone}$ in mol %	$\bar{V}_{acetone}$ in 10^{-6} m^3 mol^{-1}	$\bar{V}_{water}$
19.1	70.0	17.6

Table F.3: Measured diffusion coefficient for the binary system acetone + water at $x_{acetone}$ = 19.1 mol % and 25 °C from microfluidic experiments in this work with standard deviation $\sigma(\boldsymbol{D}^V)$. The quality of fit for the diffusion coefficient is given as RMSE$_{fit}$.

	$\boldsymbol{D}^V$ in 10^{-9} m^2 s^{-1}	RMSE$_{fit}$ in mol %
V1	0.670	0.075
V2	0.825	0.074
V3	0.646	0.070
$n_{exp} = 1$	0.714(97)	0.073
$n_{exp} = 3$	0.700	0.074
Tyn and Calus[218] at x_1 =20 mol %	0.665(17)	

Appendix G

Results: Cyclohexane + Methanol

In the following, the concentrations of the inlet solutions of cyclohexane + methanol (Table G.1) are presented together with the calculation of the partial molar volumes (Table G.2), and the determined diffusion coefficients (Table G.3).

Table G.1: Composition of the experiments with cyclohexane + methanol in mol %

	$\bar{x}$	x_l	x_r
cyclohexane	1	99.5	98.5
methanol	99	0.5	1.5
cyclohexane	5	4.5	5.5
methanol	95	95.5	94.5

The excess volume of cyclohexane (1) + methanol is described by the Redlich-Kister type function

$$V_E/(\mathrm{cm^3\,mol^{-1}}) = x_1x_2(2.008 + -0.666(x_1 - x_2) + 2.442(x_1 - x_2)^2) \qquad (G.1)$$

based on data in ref.[205] This Redlich-Kister type function results in the partial molar volumes given in Table G.2.

Diffusion coefficients for all diffusion experiments of cyclohexane (1) + methanol are given in Table G.3. The table includes evaluations with $n_{exp} = 1$ as well as $n_{exp} = 5$ and $n_{exp} = 4$, respectively.

Table G.2: Partial molar volumes of cyclohexane + methanol

$x_{\text{cyclohexane}}$ in mol %	$\bar{V}_{\text{cyclohexane}}$ in $10^{-6}\,\text{m}^3\,\text{mol}^{-1}$	$\bar{V}_{\text{methanol}}$ in $10^{-6}\,\text{m}^3\,\text{mol}^{-1}$
1	113.60	40.74
5	112.487	40.78

Table G.3: Measured diffusion coefficient for the binary system cyclohexane + methanol at $x_{\text{cyclohexane}}$ =1 mol % and 5 mol % and 25 °C from microfluidic experiments in this work with standard deviation $\sigma(\boldsymbol{D}^V)$. The quality of fit for the diffusion coefficient is given as RMSE_{fit}.

x_1 in mol %		$\boldsymbol{D}^V$ in $10^{-9}\,\text{m}^2\,\text{s}^{-1}$	RMSE_{fit} in mol %
1	V1	2.781	0.014
	V2	2.550	0.022
	V3	2.635	0.015
	V4	2.825	0.031
	V5	2.775	0.014
	$n_{\text{exp}} = 1$	2.713(116)	0.19
	$n_{\text{exp}} = 5$	2.685	0.021
5	V6	1.871	0.027
	V7	1.859	0.028
	V8	1.888	0.026
	$n_{\text{exp}} = 1$	1.873(15)	0.027
	$n_{\text{exp}} = 4$	1.876	0.027

Appendix H

Results: 1-Propanol + 1-Chlorobutane + Heptane

For 1-propanol (1) + 1-chlorobutane (2) + heptane, the composition of the experiments is given in Table H.1. Densities and viscosities of the system 1-propanol + 1-chlorobutane + n-heptane are given in Table H.2 and Table H.3. The diffusion coefficient matrix at $\bar{x}_1 = \bar{x}_2 = 33.3\,\mathrm{mol}\,\%$ is given in Table H.7. Results of individual evaluations are given in Table H.4 for $\Delta x_1 = 6.12\,\mathrm{mol}\,\%$ and $\Delta x_2 = 10.58\,\mathrm{mol}\,\%$, in Table H.5 for $\Delta x_1 = \Delta x_2 = 5\,\mathrm{mol}\,\%$, and in Table H.6 for $\Delta x_1 = 3.06\,\mathrm{mol}\,\%$ and $\Delta x_2 = 4.72\,\mathrm{mol}\,\%$.

The excess volumes determined from the densities written in Table H.2 are of the same order as the measurement accuracy of the density measurements, see Sec. 3.4. The difference between partial molar volumes and pure component molar volumes is less than 0.4 %, Therefore, the pure component molar volumes are used in the parameter estimation procedure, see Table 3.1.

Table H.1: Composition of the experiments with 1-propanol (1) + 1-chlorobutane (2) + heptane in mol %

		series A		series B	
	$\bar{x}$	x_l	x_r	x_l	x_r
1-propanol	33.3	36.36	30.24	36.36	30.24
1-chlorobutane	33.3	38.02	28.58	28.58	38.02
heptane	33.4	25.63	41.18	35.06	31.74
1-propanol	33.3	35.8	30.8	35.8	30.8
1-chlorobutane	33.3	35.8	30.8	30.8	35.8
heptane	33.4	28.4	38.4	33.4	33.4
1-propanol	33.3	34.8	31.8	34.8	31.8
1-chlorobutane	33.3	35.7	30.9	30.9	35.7
heptane	33.4	29.5	37.3	34.2	32.6

Table H.2: Densities and viscosities of the system 1-propanol (1) + 1-chlorobutane (2) + n-heptane at 25 °C

x_1	x_2	density ρ in g cm^{-3}	viscosity η in mPa s	variance of η in %
0.3580	0.3581	0.7787	0.534	1.10
0.3081	0.3080	0.76006	0.497	0.37
0.3579	0.3080	0.76688	0.531	0.86
0.3080	0.3580	0.77115	0.511	4.14

Table H.3: Densities and viscosities of the system 1-propanol (1) + 1-chlorobutane (2) + n-heptane at x1=0.3330 x2=0.3331 from 20 °C to 30 °C

T in °C	density ρ in $g\,cm^{-3}$	viscosity η in mPa s	variance of η in %
20	0.77387	0.548	0.25
24	0.77005	0.527	0.51
25	0.76907	0.522	0.32
26	0.76812	0.52	0.57
30	0.76424	0.499	0.60

Table H.4: Measured diffusion coefficient matrices for the ternary system 1-propanol (1) + 1-chlorobutane (2) + heptane at $x_1 = x_2 = 33.33\,\text{mol}\,\%$, $\Delta x_1 = 6.12\,\text{mol}\,\%$, $\Delta x_2 = 10.58\,\text{mol}\,\%$ and 25 °C from microfluidic experiments in this work with standard deviation $\sigma(\boldsymbol{D}^V)$. The quality of fit for the diffusion coefficient is given as RMSE_fit.

	$\boldsymbol{D}^V$				RMSE_fit
	D_{11}^V	D_{12}^V	D_{21}^V	D_{22}^V	
	in $10^{-9}\,\text{m}^2\,\text{s}^{-1}$				in mol %
V1	0.946	-0.246	0.513	2.148	0.272
V2	1.059	-0.362	0.340	2.530	0.242
V3	0.206	-1.114	0.477	2.520	0.278
V4	0.587	-0.609	0.171	2.431	0.253
V5	0.991	-0.326	0.308	2.389	0.278
V6	0.866	-0.031	0.341	2.461	0.247
V7	0.439	-0.814	0.493	2.481	0.257
V8	0.560	-0.632	0.356	2.544	0.245
V1V5	0.997	-0.327	0.344	2.392	0.277
V1V6	0.959	-0.276	0.265	2.503	0.265
V1V7	0.971	-0.293	0.372	2.350	0.266
V1V8	0.956	-0.268	0.292	2.472	0.260
V2V5	1.019	-0.308	0.393	2.461	0.262
V2V6	0.981	-0.258	0.313	2.569	0.248
V2V7	0.996	-0.273	0.425	2.419	0.250
V2V8	0.978	-0.250	0.340	2.537	0.244
V3V5	0.993	-0.331	0.313	2.380	0.279
V3V6	0.947	-0.273	0.229	2.495	0.266
V3V7	0.957	-0.280	0.336	2.366	0.270
V3V8	0.945	-0.267	0.254	2.466	0.262
V4V5	1.039	-0.271	0.358	2.426	0.265
V4V6	1.000	-0.224	0.277	2.535	0.251
V4V7	1.017	-0.239	0.388	2.385	0.253
V4V8	0.998	-0.217	0.302	2.504	0.247
$n_\text{exp} = 1$	0.707(304)	-0.517(346)	0.375(115)	2.438(128)	0.26
$n_\text{exp} = 2$	0.985(28)	-0.272(32)	0.325(55)	2.454(67)	0.26
$n_\text{exp} = 8$	0.984	-0.273	0.325	2.453	0.262

Table H.5: Measured diffusion coefficient matrices for the ternary system 1-propanol (1) + 1-chlorobutane (2) + heptane at $x_1 = x_2 = 33.33\,\mathrm{mol\,\%}$, $\Delta x_1 = 5\,\mathrm{mol\,\%}$, $\Delta x_2 = 5\,\mathrm{mol\,\%}$ and 25 °C from microfluidic experiments in this work with standard deviation $\sigma(\boldsymbol{D}^V)$. The quality of fit for the diffusion coefficient is given as $\mathrm{RMSE_{fit}}$. The results of repeated evaluations and for $n_{\mathrm{exp}} = 8$ are given in Table. H.7.

	$\boldsymbol{D}^V$				$\mathrm{RMSE_{fit}}$
	D^V_{11}	D^V_{12}	D^V_{21}	D^V_{22}	
	in $10^{-9}\,\mathrm{m^2\,s^{-1}}$				in mol %
V9	0.962	-0.113	0.347	2.574	0.247
V10	0.667	-0.923	0.158	2.278	0.263
V11	0.782	-0.545	0.224	2.454	0.236
V12	0.867	-0.447	0.432	2.818	0.218
V13	0.837	0.047	0.410	2.254	0.283
V14	0.968	-0.182	0.362	2.408	0.230
V15	0.974	-0.281	0.348	2.509	0.210
V16	0.690	-0.873	0.759	3.161	0.238
V17	0.400	-1.214	0.491	2.771	0.211
V9V13	1.052	-0.293	0.339	2.580	0.255
V9V14	1.001	-0.192	0.323	2.614	0.241
V9V15	1.018	-0.226	0.312	2.638	0.234
V9V16	1.043	-0.273	0.366	2.533	0.242
V9V17	1.023	-0.238	0.347	2.572	0.230
V10V13	1.020	-0.343	0.288	2.473	0.274
V10V14	0.971	-0.242	0.273	2.508	0.261
V10V15	0.988	-0.277	0.262	2.532	0.254
V10V16	1.012	-0.326	0.314	2.424	0.262
V10V17	0.992	-0.288	0.293	2.467	0.250
V11V13	1.034	-0.324	0.311	2.521	0.247
V11V14	0.982	-0.223	0.295	2.553	0.235
V11V15	1.000	-0.257	0.284	2.577	0.227
V11V16	1.025	-0.305	0.336	2.472	0.236
V11V17	1.005	-0.268	0.317	2.512	0.223
V12V13	1.013	-0.360	0.323	2.551	0.238
V12V14	0.961	-0.260	0.307	2.585	0.223
V12V15	0.978	-0.295	0.295	2.606	0.215
V12V16	1.005	-0.342	0.350	2.504	0.224
V12V17	0.984	-0.304	0.330	2.543	0.211

Table H.6: Measured diffusion coefficient matrices for the ternary system 1-propanol (1) + 1-chlorobutane (2) + heptane at $x_1 = x_2 = 33.33\,\mathrm{mol\,\%}$ with $\Delta x_1 = 3.06\,\mathrm{mol\,\%}$ and $\Delta x_2 = 4.72\,\mathrm{mol\,\%}$ and 25 °C from microfluidic experiments in this work with standard deviation $\sigma(\boldsymbol{D}^V)$. The quality of fit for the diffusion coefficient is given as $\mathrm{RMSE_{fit}}$.

	$\boldsymbol{D}^V$				$\mathrm{RMSE_{fit}}$
	D^V_{11}	D^V_{12}	D^V_{21}	D^V_{22}	
	in $10^{-9}\,\mathrm{m^2\,s^{-1}}$				in mol %
V18	1.013	-0.217	0.271	2.513	0.295
V19	0.919	0.010	0.334	2.472	0.245
V20	0.262	-1.001	0.463	2.594	0.250
V21	0.087	-1.162	0.540	2.783	0.237
V22	0.876	-0.185	0.373	2.275	0.298
V23	1.231	-0.549	0.216	2.608	0.247
V24	0.410	-0.819	0.212	2.342	0.241
V25	1.163	-0.048	0.307	2.416	0.245
V17V22	1.043	-0.260	0.302	2.468	0.274
V17V23	1.033	-0.249	0.237	2.560	0.268
V17V24	1.012	-0.219	0.304	2.470	0.269
V17V25	1.000	-0.195	0.327	2.436	0.271
V19V22	1.076	-0.220	0.320	2.482	0.248
V19V23	1.066	-0.208	0.255	2.574	0.242
V19V24	1.053	-0.187	0.323	2.478	0.244
V19V25	1.036	-0.155	0.346	2.450	0.245
V20V22	0.970	-0.325	0.280	2.426	0.276
V20V23	0.970	-0.325	0.280	2.426	0.276
V20V24	0.961	-0.313	0.217	2.520	0.271
V20V25	0.935	-0.285	0.278	2.437	0.272
V21V22	0.931	-0.259	0.305	2.394	0.273
V21V23	1.034	-0.279	0.315	2.484	0.249
V21V24	1.024	-0.268	0.248	2.575	0.242
V21V25	1.011	-0.246	0.317	2.479	0.245
$n_\mathrm{exp} = 1$	0.745(432)	-0.497(453)	0.340(116)	2.500(162)	0.257
$n_\mathrm{exp} = 2$	1.010(45)	-0.250(50)	0.291(36)	2.479(54)	0.260
$n_\mathrm{exp} = 8$	1.010	-0.244	0.293	2.478	0.259

Table H.7: Measured diffusion coefficient matrices for 1-propanol + 1-chlorobutane + heptane at $x_{\text{1-propanol}} = x_{\text{1-chlorobutane}} = 33.3\,\text{mol}\,\%$ with $\Delta x_1 = \Delta x_2 = 5\,\text{mol}\,\%$ and $25\,^{\circ}\text{C}$ by microfluidic experiment in this work with standard deviation $\sigma(\boldsymbol{D}^V)$ from repeated experiments in comparison to Käshammer *et al.*[2]. The quality of fit for the diffusion coefficient is given as RMSE_{fit}.

	D_{11}	D_{12}	D_{21}	D_{22}	RMSE_{fit}
		in $1 \times 10^{-9}\,\text{m}^2\,\text{s}^{-1}$			in mol %
$n_{\text{exp}} = 1$[a]	0.794(188)	−0.504(422)	0.392(171)	2.581(291)	0.283
$n_{\text{exp}} = 2$[b]	1.005(24)	−0.282(45)	0.313(27)	2.538(55)	0.284
$n_{\text{exp}} = 9$[c]	1.004	−0.282	0.313	2.540	0.284
literature	1.033(16)	−0.174(25)	0.255(12)	2.46(17)	

[a] estimated from nine experiments in nine separate evaluations

[b] estimated from all combinations of two experiments, one from series A and series B each

[c] estimated from all experiments simultaneously, four from series A and five from series B each. No standard deviations from repeated experiments are available, but standard deviations are expected to be less than for $n_{\text{exp}} = 2$ as more data was used.

Appendix I

Results: Cyclohexane + Toluene + Methanol

For cyclohexane (1) + toluene (2) + methanol, the composition of the experiments is given in Table I.1. Measured densities and viscosities are given in Tables I.2-I.4. The resulting partial molar volumes are given in Table I.5 based on a Redlich-Kister type function (I.1)-(I.5). The diffusion coefficient matrix at $\bar{x}_1 = \bar{x}_2 = 5\,\mathrm{mol}\,\%$ is given in Table I.6 and at $\bar{x}_1 = \bar{x}_2 = 10\,\mathrm{mol}\,\%$ in Table I.7.

Table I.1: Composition of the experiments with cyclohexane + toluene + methanol in mol %

		series A		series B	
	$\bar{x}$	x_l	x_r	x_l	x_r
cyclohexane	5	5.5	4.5	5.5	4.5
toluene	5	5.5	4.5	4.5	5.5
methanol	90	89	91	90	90
cyclohexane	10	11	9	9	11
toluene	10	11	9	11	9
methanol	80	78	82	80	80

A Redlich-Kister type function with binary interactions is used to describe the

Table I.2: Densities and viscosities of the system cyclohexane (1) + toluene (2) + methanol at 25 °C

x_1	x_2	density ρ in $g\,cm^{-3}$	viscosity η in mPa s	variance of η in %
0.1386	0.1306	0.7965	0.5950	0.03
0.0895	0.0912	0.7942	0.5817	0.01
0.0878	0.0957	0.7949	0.5833	0.57
0.1031	0.0933	0.7934	0.5885	0.52
0.0564	0.0588	0.7914	0.5687	0.02
0.0451	0.0451	0.7904	0.5678	0.55
0.0456	0.0438	0.7901	0.565	0.02
0.0548	0.0452	0.7895	0.5672	0.08

Table I.3: Densities and viscosities of the system cyclohexane (1) + toluene (2) + methanol at $x_1 = 9.85$ mol % and $x_2 = 11.32$ mol % from 20 °C to 30 °C

T in °C	density ρ in $g\,cm^{-3}$	viscosity η in mPa s	variance of η in %
20	0.80151	0.6294	0.69
24	0.79767	0.5935	0.4
25	0.79675	0.5853	0.01
26	0.79576	0.5773	0.04
30	0.78188	0.5468	0.05

Table I.4: Densities and viscosities of the system cyclohexane (1) + toluene (2) + methanol at $x_1 = 5.21$ mol % and $x_2 = 5.01$ mol % from 20 °C to 30 °C

T in °C	density ρ in g cm^{-3}	viscosity η in mPa s	variance of η in %
20	0.79537	0.6075	0.02
24	0.79157	0.5748	0.03
25	0.7906	0.5669	0.01
26	0.78965	0.5595	0.01
30	0.78583	0.5305	0.01

excess volume

$$V_{\mathrm{E}} = x_1 x_2(1.93E - 05) + x_1 x_3(2.82E - 06) + x_2 x_3(-1.91E - 07) \tag{I.1}$$

at 5 mol % and

$$V_{\mathrm{E}} = x_1 x_2(3.08E - 06) + x_1 x_2(1.82E - 07) + x_2 x_3(2.82E - 06) \tag{I.2}$$

at 10 mol % based on data in Table I.2.

The resulting partial molar volumes are

$$\bar{V}_1 = A_1 x_2^2 + x_3(A_1 + A_2 - A_3)x_2 + A_2 x_3^2 + V_1 \;, \tag{I.3}$$

$$\bar{V}_2 = A_1 x_1^2 + x_3(A_1 - A_2 + A_3)x_1 + A_3 x_3^2 + V_2 \text{ and} \tag{I.4}$$

$$\bar{V}_3 = A_2 x_1^2 - x_2(A_1 - A_2 - A_3)x_1 + A_3 x_3^2 + V_3 \;. \tag{I.5}$$

Table I.5: Partial molar volumes of cyclohexane + toluene + methanol

$x_{cyclohexane} = x_{toluene}$ in mol %	$\bar{V}_{cyclohexane}$	$\bar{V}_{toluene}$	$\bar{V}_{methanol}$
	in $10^{-6}\,m^3\,mol^{-1}$		
5	112.10	107.48	40.71
10	111.31	107.07	40.75

Table I.6: Measured diffusion coefficient matrices for cyclohexane (1) + toluene (2) + methanol at $x_1 = x_2 = 5$ mol % and 298.15 K estimated from eqn (3.10). Standard deviations are given for repeated evaluations in comparison to Großmann *et al.*[1] For the diffusion coefficients measured in this work, also the $RMSE_{fit}$ is given.

	D_{11}	D_{12}	D_{21}	D_{22}	$RMSE_{fit}$
		in $1 \times 10^{-9}\,m^2\,s^{-1}$			in mol %
V1	1.509	-0.404	-0.395	2.072	0.048
V2	1.702	-0.161	-0.460	2.035	0.108
V3	1.340	-0.635	-0.750	2.108	0.050
V4	1.172	-0.528	-0.680	2.252	0.046
V5	0.411	0.436	-1.255	2.249	0.080
V6	1.036	-0.252	-0.819	2.328	0.052
V7	1.660	-0.102	-0.737	1.641	0.058
V8	1.724	0.000	-1.186	1.181	0.074
V9	1.470	-0.483	-0.561	1.982	0.052
V10	1.512	-0.332	-0.282	2.090	0.053
V11	1.655	-0.160	-0.415	2.023	0.054
V12	1.806	-1.271	-0.563	1.988	0.070
V13	0.151	1.039	-2.075	3.890	0.061
V14	1.680	-1.183	-0.497	1.884	0.052
V1V3	1.319	-0.625	-0.509	1.943	0.050

Table I.6: Measured diffusion coefficient matrices for cyclohexane (1) + toluene (2) + methanol at $x_1 = x_2 = 5\,\mathrm{mol\,\%}$ and 298.15 K, continued.

	D_{11}	D_{12}	D_{21}	D_{22}	$\mathrm{RMSE_{fit}}$
		in $1 \times 10^{-9}\,\mathrm{m^2\,s^{-1}}$			in mol %
V1V4	1.287	-0.648	-0.483	1.969	0.047
V1V5	1.327	-0.616	-0.485	1.970	0.059
V1V6	1.335	-0.608	-0.502	1.949	0.050
V1V12	1.298	-0.641	-0.525	1.923	0.060
V1V13	1.412	-0.504	-0.494	1.948	0.055
V1V14	1.273	-0.678	-0.527	1.920	0.050
V2V3	1.304	-0.613	-0.521	1.956	0.085
V2V4	1.278	-0.652	-0.482	2.008	0.083
V2V5	1.317	-0.606	-0.497	1.990	0.091
V2V6	1.322	-0.596	-0.515	1.966	0.086
V2V12	1.290	-0.639	-0.526	1.955	0.092
V2V13	1.409	-0.495	-0.498	1.986	0.088
V2V14	1.265	-0.668	-0.539	1.939	0.085
V7V3	1.244	-0.569	-0.486	1.914	0.054
V7V4	1.235	-0.579	-0.466	1.937	0.052
V7V5	1.269	-0.548	-0.465	1.947	0.063
V7V6	1.277	-0.538	-0.483	1.924	0.055
V7V12	1.243	-0.578	-0.498	1.905	0.064
V7V13	1.350	-0.440	-0.477	1.911	0.059
V7V14	1.218	-0.608	-0.508	1.894	0.055
V8V3	1.324	-0.624	-0.559	1.981	0.062
V8V4	1.290	-0.662	-0.510	2.036	0.060
V8V5	1.328	-0.614	-0.526	2.018	0.071
V8V6	1.335	-0.607	-0.545	1.996	0.063
V8V12	1.302	-0.648	-0.552	1.983	0.072

Table I.6: Measured diffusion coefficient matrices for cyclohexane (1) + toluene (2) + methanol at $x_1 = x_2 = 5\,\mathrm{mol\,\%}$ and 298.15 K, continued.

	D_{11}	D_{12}	D_{21}	D_{22}	$\mathrm{RMSE_{fit}}$
		in $1 \times 10^{-9}\,\mathrm{m^2\,s^{-1}}$			in mol %
V8V13	1.419	-0.502	-0.528	2.017	0.067
V8V14	1.277	-0.678	-0.566	1.966	0.063
V9V3	1.341	-0.643	-0.556	1.979	0.051
V9V4	1.307	-0.686	-0.510	2.035	0.049
V9V5	1.346	-0.637	-0.526	2.017	0.061
V9V6	1.352	-0.631	-0.543	1.994	0.052
V9V12	1.318	-0.673	-0.552	1.982	0.062
V9V13	1.435	-0.526	-0.526	2.015	0.057
V9V14	1.295	-0.703	-0.566	1.964	0.052
V10V3	1.282	-0.599	-0.465	1.901	0.051
V10V4	1.259	-0.619	-0.437	1.926	0.050
V10V5	1.298	-0.582	-0.442	1.922	0.061
V10V6	1.305	-0.574	-0.461	1.899	0.053
V10V12	1.270	-0.609	-0.482	1.876	0.062
V10V13	1.384	-0.470	-0.452	1.901	0.057
V10V14	1.246	-0.644	-0.485	1.869	0.053
V11V3	1.278	-0.592	-0.469	1.914	0.052
V11V4	1.253	-0.622	-0.436	1.952	0.050
V11V5	1.292	-0.576	-0.451	1.935	0.061
V11V6	1.299	-0.568	-0.470	1.912	0.053
V11V12	1.265	-0.608	-0.480	1.899	0.062
V11V13	1.385	-0.466	-0.453	1.932	0.057
V11V14	1.241	-0.638	-0.494	1.883	0.053

Table I.6: Measured diffusion coefficient matrices for cyclohexane (1) + toluene (2) + methanol at $x_1 = x_2 = 5\,\text{mol}\,\%$ and 298.15 K, continued.

	D_{11}	D_{12}	D_{21}	D_{22}	RMSE_fit
		in $1 \times 10^{-9}\,\text{m}^2\,\text{s}^{-1}$			in mol %
$n_\text{exp} = 1^\text{a}$	1.193(493)	−0.401(929)	−0.396(468)	2.248(505)	0.033
$n_\text{exp} = 2^\text{b}$	1.773(58)	−0.467(65)	−0.366(27)	2.146(30)	0.033
$n_\text{exp} = 10^\text{c}$	1.777	−0.472	−0.368	2.150	0.033
literature[1]	1.651(6)	−0.101(9)	−0.381(16)	1.805(10)	
literature[1]	1.721(8)	−0.031(9)	−0.364(18)	1.855(11)	

Table I.7: Measured diffusion coefficient matrices for cyclohexane (1) + toluene (2) + methanol at $x_1 = x_2 = 10\,\text{mol}\,\%$ and 298.15 K estimated from eqn (3.10) with standard deviations from repeated experiments in comparison to Großmann *et al.*[1] For the diffusion coefficients measured in this work, also the RMSE_fit is given.

	D_{11}	D_{12}	D_{21}	D_{22}	RMSE_fit
		in $1 \times 10^{-9}\,\text{m}^2\,\text{s}^{-1}$			in mol %
V1	1.509	-0.404	-0.395	2.072	0.048
V2	1.702	-0.161	-0.460	2.035	0.108
V3	1.340	-0.635	-0.750	2.108	0.050
V4	1.172	-0.528	-0.680	2.252	0.046
V5	0.411	0.436	-1.255	2.249	0.080
V6	1.036	-0.252	-0.819	2.328	0.052
V7	1.660	-0.102	-0.737	1.641	0.058
V8	1.724	0.000	-1.186	1.181	0.074

[a] estimated from separate experiments in separate evaluations

[b] estimated from all combinations of two independent experiments

[c] estimated from a total of ten experiments using series A and B. No standard deviations from repeated evaluations are available, but standard deviations are expected to be less than for $n_\text{exp} = 3$ as more data was used.

Table I.7: Measured diffusion coefficient matrices for cyclohexane (1) + toluene (2) + methanol at $x_1 = x_2 = 10\,\text{mol}\,\%$ and 298.15 K, continued.

	D_{11}	D_{12}	D_{21}	D_{22}	RMSE_{fit}
		in $1 \times 10^{-9}\,\text{m}^2\,\text{s}^{-1}$			in mol %
V9	1.470	-0.483	-0.561	1.982	0.052
V10	1.512	-0.332	-0.282	2.090	0.053
V11	1.655	-0.160	-0.415	2.023	0.054
V12	1.806	-1.271	-0.563	1.988	0.070
V13	0.151	1.039	-2.075	3.890	0.061
V14	1.680	-1.183	-0.497	1.884	0.052
V1V3	1.319	-0.625	-0.509	1.943	0.050
V1V4	1.287	-0.648	-0.483	1.969	0.047
V1V5	1.327	-0.616	-0.485	1.970	0.059
V1V6	1.335	-0.608	-0.502	1.949	0.050
V1V12	1.298	-0.641	-0.525	1.923	0.060
V1V13	1.412	-0.504	-0.494	1.948	0.055
V1V14	1.273	-0.678	-0.527	1.920	0.050
V2V3	1.304	-0.613	-0.521	1.956	0.085
V2V4	1.278	-0.652	-0.482	2.008	0.083
V2V5	1.317	-0.606	-0.497	1.990	0.091
V2V6	1.322	-0.596	-0.515	1.966	0.086
V2V12	1.290	-0.639	-0.526	1.955	0.092
V2V13	1.409	-0.495	-0.498	1.986	0.088
V2V14	1.265	-0.668	-0.539	1.939	0.085
V7V3	1.244	-0.569	-0.486	1.914	0.054
V7V4	1.235	-0.579	-0.466	1.937	0.052
V7V5	1.269	-0.548	-0.465	1.947	0.063
V7V6	1.277	-0.538	-0.483	1.924	0.055
V7V12	1.243	-0.578	-0.498	1.905	0.064

Table I.7: Measured diffusion coefficient matrices for cyclohexane (1) + toluene (2) + methanol at $x_1 = x_2 = 10\,\mathrm{mol\,\%}$ and 298.15 K, continued.

	D_{11}	D_{12}	D_{21}	D_{22}	$\mathrm{RMSE_{fit}}$
		in $1 \times 10^{-9}\,\mathrm{m^2\,s^{-1}}$			in mol %
V7V13	1.350	-0.440	-0.477	1.911	0.059
V7V14	1.218	-0.608	-0.508	1.894	0.055
V8V3	1.324	-0.624	-0.559	1.981	0.062
V8V4	1.290	-0.662	-0.510	2.036	0.060
V8V5	1.328	-0.614	-0.526	2.018	0.071
V8V6	1.335	-0.607	-0.545	1.996	0.063
V8V12	1.302	-0.648	-0.552	1.983	0.072
V8V13	1.419	-0.502	-0.528	2.017	0.067
V8V14	1.277	-0.678	-0.566	1.966	0.063
V9V3	1.341	-0.643	-0.556	1.979	0.051
V9V4	1.307	-0.686	-0.510	2.035	0.049
V9V5	1.346	-0.637	-0.526	2.017	0.061
V9V6	1.352	-0.631	-0.543	1.994	0.052
V9V12	1.318	-0.673	-0.552	1.982	0.062
V9V13	1.435	-0.526	-0.526	2.015	0.057
V9V14	1.295	-0.703	-0.566	1.964	0.052
V10V3	1.282	-0.599	-0.465	1.901	0.051
V10V4	1.259	-0.619	-0.437	1.926	0.050
V10V5	1.298	-0.582	-0.442	1.922	0.061
V10V6	1.305	-0.574	-0.461	1.899	0.053
V10V12	1.270	-0.609	-0.482	1.876	0.062
V10V13	1.384	-0.470	-0.452	1.901	0.057
V10V14	1.246	-0.644	-0.485	1.869	0.053
V11V3	1.278	-0.592	-0.469	1.914	0.052
V11V4	1.253	-0.622	-0.436	1.952	0.050

Table I.7: Measured diffusion coefficient matrices for cyclohexane (1) + toluene (2) + methanol at $x_1 = x_2 = 10\,\mathrm{mol\,\%}$ and 298.15 K, continued.

	D_{11}	D_{12}	D_{21}	D_{22}	$\mathrm{RMSE_{fit}}$
		in $1 \times 10^{-9}\,\mathrm{m^2\,s^{-1}}$			in mol %
V11V5	1.292	-0.576	-0.451	1.935	0.061
V11V6	1.299	-0.568	-0.470	1.912	0.053
V11V12	1.265	-0.608	-0.480	1.899	0.062
V11V13	1.385	-0.466	-0.453	1.932	0.057
V11V14	1.241	-0.638	-0.494	1.883	0.053
$n_{\mathrm{exp}} = 1^{\mathrm{d}}$	1.345(484)	−0.288(563)	−0.763(454)	2.123(563)	0.061
$n_{\mathrm{exp}} = 2^{\mathrm{e}}$	1.306(50)	−0.601(60)	−0.501(35)	1.950(43)	0.061
$n_{\mathrm{exp}} = 14^{\mathrm{f}}$	1.307	−0.604	−0.495	1.952	0.063
literature[1]	1.228(45)	−0.411(29)	−0.507(83)	1.653(48)	
literature[1]	1.270(50)	−0.428(32)	−0.559(89)	1.681(52)	

[d] estimated from separate experiments in separate evaluations

[e] estimated from all combinations of two independent experiments

[f] estimated from a total of 14 experiments using series A and B. No standard deviations from repeated evaluations are available, but standard deviations are expected to be less than for $n_{\mathrm{exp}} = 2$ as more data was used.

Appendix J

Results: Water + Acetone + Toluene

For water (1) + acetone (2) + toluene, the composition of the experiments is given in Table J.1. Densities and viscosities are given in Table J.2. The diffusion coefficient matrices for all four concentrations are given in Tables J.3 to J.6.

Table J.1: Composition of the experiments with water + acetone + toluene in mol %

		series A		series B	
	$\bar{x}$	x_l	x_r	x_l	x_r
water	85.17	85.29	85.05	85.52	84.84
acetone	14.74	14.69	14.79	14.40	15.08
toluene	0.09	0.00	0.16	0.08	0.08
water	80.79	80.85	80.73	81.5	80.08
acetone	19.07	19.15	18.99	18.38	19.76
toluene	0.14	0.00	0.28	0.12	0.16
water	70.28	70.58	69.98	70.58	69.98
acetone	29.07	29.22	28.92	28.62	29.52
toluene	0.65	0.20	1.10	0.80	0.50
water	62.95	64.45	61.45	63.08	62.81
acetone	35.60	34.50	36.70	35.98	35.24
toluene	1.45	1.05	1.85	0.95	1.95

Table J.2: Measured densities and viscosities for water (1) + acetone (2) + toluene at 298.15 K.

x_1 wt%	x_2 wt%	ρ g cm^{-3}	η mPa s	$u_r(\eta)$ %
85.29	14.69	0.9436	1.342	0.13
85.05	14.79	0.9427	1.421	0.69
85.52	14.40	0.9446	1.359	0.10
84.84	15.08	0.9421	1.354	0.03
80.85	19.15	0.9297	1.313	0.05
80.73	18.99	0.9292	1.318	0.36
81.50	18.38	0.9312	1.335	0.48
80.08	19.76	0.9265	1.296	0.49
70.58	29.22	0.8987	1.096	0.63
70.13	28.76	0.8976	1.091	0.52
70.55	28.67	0.8995	1.094	0.03
69.97	29.53	0.8975	1.077	0.01
64.39	34.55	0.8835	0.953	0.01
61.45	36.70	0.8768	0.898	0.01
63.10	35.94	0.8806	0.925	0.02
62.85	35.23	0.8799	0.942	0.56

Table J.3: Measured diffusion coefficient matrices for water (1) + acetone (2) + toluene at x_1 =85.17 mol %, x_2 =14.74 mol %, and 298.15 K estimated from eqn (3.10) with standard deviations from repeated experiments. The quality of fit is given as RMSE$_{\text{fit}}$.

	D_{11}	D_{12}	D_{21}	D_{22}	RMSE$_{\text{fit}}$
		in $1 \times 10^{-9}\,\text{m}^2\,\text{s}^{-1}$			in mol %
V1	56.452	324.931	-9.464	-54.460	0.102

Table J.3: Measured diffusion coefficient matrices for water (1) + acetone (2) + toluene at x_1 =85.17 mol %, x_2 =14.74 mol %, and 298.15 K, continued.

	D_{11}	D_{12}	D_{21}	D_{22}	$RMSE_{fit}$
		in $1 \times 10^{-9}\,m^2\,s^{-1}$			in mol %
V2	78.132	390.117	-8.672	-43.188	0.091
V3	3.829	32.881	-0.324	-2.703	0.058
V4	-3.046	-11.083	1.928	6.866	0.051
V5	-71.909	-306.741	27.060	114.994	0.048
V6	-0.679	-6.256	0.205	1.746	0.063
V7	-644.493	-2704.228	158.643	665.629	0.054
V8	-218.389	-917.854	54.260	228.027	0.282
V1 V5	-2.772	-14.234	0.861	4.260	0.090
V1 V6	-2.428	-12.253	0.822	3.925	0.092
V1 V7	-3.400	-17.539	0.988	4.978	0.089
V1 V8	-2.394	-12.066	0.820	3.984	0.090
V2 V5	55.538	249.706	-11.369	-51.070	0.000
V2 V6	-1.849	-10.792	0.593	3.250	0.079
V2 V7	-1.895	-11.154	0.677	3.707	0.075
V2 V8	-1.214	-7.580	0.519	2.896	0.076
V3 V5	46.353	213.089	-9.738	-44.750	0.000
V3 V6	0.596	1.383	-0.019	0.919	0.067
V3 V7	-3.328	-17.234	0.972	4.912	0.056
V3 V8	-2.770	-13.356	0.890	4.126	0.058
V4 V5	84.870	292.724	-24.210	-83.498	0.000
V4 V6	43.622	122.674	-15.081	-42.405	0.000
V4 V7	115.813	394.298	-33.558	-114.250	0.000
V4 V8	81.861	210.663	-31.234	-80.373	0.000

Table J.3: Measured diffusion coefficient matrices for water (1) + acetone (2) + toluene at x_1 =85.17 mol %, x_2 =14.74 mol %, and 298.15 K, continued.

	D_{11}	D_{12}	D_{21}	D_{22}	$\mathrm{RMSE_{fit}}$
		in $1 \times 10^{-9}\,\mathrm{m^2\,s^{-1}}$			in mol %
$n_{\mathrm{exp}} = 1^{\mathrm{a}}$	125.5(2454)	−512.6(10422)	33.2(595)	137.2(2415)	0.093
$n_{\mathrm{exp}} = 2^{\mathrm{b}}$	−2.145(1185)	−11.483(5407)	−0.712(300)	3.676(1170)	0.073
$n_{\mathrm{exp}} = 7^{\mathrm{c}}$	−2.824	−14.283	0.995	4.728	0.073

Table J.4: Measured diffusion coefficient matrices for water (1) + acetone (2) + toluene at x_1 =80.79 mol %, x_2 =19.07 mol %, and 298.15 K estimated from eqn (3.10) with standard deviations from repeated experiments. The quality of fit is given as $\mathrm{RMSE_{fit}}$.

	D_{11}	D_{12}	D_{21}	D_{22}	$\mathrm{RMSE_{fit}}$
		in $1 \times 10^{-9}\,\mathrm{m^2\,s^{-1}}$			in mol %
V1	-2.290	-12.834	0.884	4.544	0.078
V2	233.4	2327.8	-22.4	-223.3	0.072
V3	-1.741	-8.960	0.652	3.092	0.076
V4	23.638	129.658	-4.033	-22.095	0.060
V5	-12.667	-54.676	4.767	20.279	0.075
V6	-5.081	-20.744	6.060	22.810	0.180
V1 V4	26.64	265.39	-2.541	-25.29	0.074
V1 V5	-2.220	-11.693	0.954	4.605	0.076
V1 V6	-2.265	-12.058	0.984	4.696	0.140
V2 V4	-1.575	-8.418	0.647	3.137	0.058
V2 V5	170.17	720.26	-39.88	-168.79	0.070

[a] estimated from all evaluations of individual experiments

[b] estimated from combinations of two independent experiments. The determinants of all diffusion coefficient matrices are in the order of 10^{-19}. To avoid non-positive definite diffusion coefficient matrices, the six combinations with $D_{ij} > 10^{-8}\,\mathrm{m^2\,s^{-1}}$ were excluded.

[c] estimated from seven experiments simultaneously, experiment 2 was excluded.

Table J.4: Measured diffusion coefficient matrices for water (1) + acetone (2) + toluene at x_1 =85.17 mol %, x_2 =14.74 mol %, and 298.15 K, continued.

	D_{11}	D_{12}	D_{21}	D_{22}	$\mathrm{RMSE_{fit}}$
			in $1 \times 10^{-9}\,\mathrm{m^2\,s^{-1}}$		in mol %
V2 V6	-1.768	-9.983	0.587	3.090	0.136
V3 V4	-1.485	-7.894	0.654	3.060	0.067
V3 V5	-1.286	-7.772	0.586	3.036	0.076
V3 V6	-1.245	-7.702	0.612	3.217	0.139
$n_{\mathrm{exp}} = 1$[d]	39.20(9591)	393.37(94981)	−2.34(1044)	−32.44(9484)	0.090
$n_{\mathrm{exp}} = 2$[e]	−1.449(148)	−8.028(343)	0.629(38)	3.078(53)	0.070
$n_{\mathrm{exp}} = 5$[f]	−1.759	−9.693	0.711	3.549	0.069

Table J.5: Measured diffusion coefficient matrices for water (1) + acetone (2) + toluene at x_1 =70.28 mol %, x_2 =29.07 mol %, and 298.15 K estimated from eqn (3.10) with standard deviations from repeated experiments. The quality of fit is given as $\mathrm{RMSE_{fit}}$.

	D_{11}	D_{12}	D_{21}	D_{22}	$\mathrm{RMSE_{fit}}$
			in $1 \times 10^{-9}\,\mathrm{m^2\,s^{-1}}$		in mol %
V1	-0.641	-5.830	0.411	2.584	0.112
V2	-0.267	-3.136	0.315	1.914	0.110
V3	-0.491	-4.692	0.381	2.394	0.109
V4	0.426	-5.198	0.232	2.625	0.110
V5	-0.011	-3.650	0.202	1.705	0.119
V6	-0.369	-5.564	0.363	2.471	0.120

[d] estimated from all evaluations of individual experiments

[e] estimated from combinations of two independent experiments. All combinations with experiment V6 were excluded due to a large $\mathrm{RMSE_{fit}}$. V2 V5 was excluded as the diffusion coefficient matrix was two orders of magnitudes larger than all other matrices.

[f] estimated from five experiments simultaneously, experiment 6 was excluded.

Table J.5: Measured diffusion coefficient matrix for water (1) + acetone (2) + toluene at x_1 =70.28 mol %, x_2 =29.07 mol %, and 298.15 K, continued.

	D_{11}	D_{12}	D_{21}	D_{22}	RMSE$_{fit}$
		in $1 \times 10^{-9}\,m^2\,s^{-1}$			in mol %
V1 V4	-0.636	-6.033	0.436	2.864	0.112
V1 V5	-0.557	-4.684	0.410	2.215	0.116
V1 V6	-0.621	-5.655	0.419	2.594	0.116
V2 V4	-0.358	-4.814	0.317	2.584	0.111
V2 V5	-0.291	-3.426	0.308	1.932	0.114
V2 V6	-0.355	-4.461	0.307	2.301	0.115
V3 V4	-0.521	-5.414	0.386	2.731	0.110
V3 V5	-0.461	-4.202	0.380	2.190	0.114
V3 V6	-0.513	-5.064	0.379	2.502	0.115
$n_{exp} = 1$[g]	-0.225(384)	-4.678(1079)	0.317(84)	2.282(381)	0.113
$n_{exp} = 2$[h]	-0.470(152)	-4.845(923)	0.366(61)	2.415(331)	0.114
$n_{exp} = 6$[i]	-0.477	-4.956	0.365	2.458	0.114

Table J.6: Measured diffusion coefficient matrix for water (1) + acetone (2) + toluene at x_1 =62.95 mol %, x_2 =35.60 mol %, and 298.15 K estimated from eqn (3.10) with standard deviations from repeated experiments. The quality of fit is given as RMSE$_{fit}$.

	D_{11}	D_{12}	D_{21}	D_{22}	RMSE$_{fit}$
		in $1 \times 10^{-9}\,m^2\,s^{-1}$			in mol %
V1	-0.756	-6.879	0.512	3.660	0.114
V2	68.440	508.652	-8.890	-66.061	0.138
V3	-0.621	-6.241	0.480	3.378	0.102

[g]estimated from all evaluations of individual experiments
[h]estimated from combinations of two independent experiments.
[i]estimated from five experiments simultaneously.

Table J.6: Measured diffusion coefficient matrix for water (1) + acetone (2) + toluene at x_1 =62.95 mol %, x_2 =35.60 mol %, and 298.15 K, continued.

	D_{11}	D_{12}	D_{21}	D_{22}	$RMSE_{fit}$
		in $1 \times 10^{-9}\,m^2\,s^{-1}$			in mol %
V4	-0.221	-3.032	0.307	2.078	0.107
V5	-525.7	3030.8	-91.5	527.6	0.175
V6	-0.062	-6.872	0.103	3.473	0.251
V1 V4	-0.243	-3.026	0.337	2.103	0.110
V1 V5	-0.343	-3.843	0.393	2.561	0.113
V1 V6	-0.516	-5.280	0.484	3.506	0.210
V2 V4	-0.215	-3.165	0.300	2.059	0.123
V2 V5	-0.316	-3.982	0.374	2.620	0.125
V2 V6	-0.462	-5.123	0.402	3.024	0.214
V3 V4	-0.214	-3.311	0.416	2.810	0.107
V3 V5	-0.329	-3.995	0.384	2.500	0.107
V3 V6	-0.453	-5.078	0.367	2.710	0.205
$n_{exp} = 1$[j]	-76.4(2217)	586.0(12152)	-16.5(369)	79.0(2215)	0.148
$n_{exp} = 2$[k]	-0.315(112)	-4.089(876)	0.384(51)	2.655(444)	0.143
$n_{exp} = 6$[l]	-0.359	-4.170	0.373	2.567	0.154

[j]estimated from all evaluations of individual experiments

[k]estimated from combinations of two independent experiments.

[l]estimated from five experiments simultaneously.

Appendix K

Results: Cyclohexane + Toluene + Acetone + Methanol

For cyclohexane (1) + toluene (2) + acetone (3) + methanol, the composition of the diffusion experiments is given in Table K.1. The densities and viscosities of the quaternary system are given in Tables K.3 and K.2. The diffusion coefficient matrix was determined using partial molar volumes. The partial molar volumes are given in Table K.4. Diffusion coefficient matrices determined from eqn (3.10) are given in Table K.5. The diffusion coefficient matrix at $\bar{x}_1 = \bar{x}_2 = \bar{x}_3 = 5\,\mathrm{mol}\,\%$ is given in Table K.6 averaged over repeated evaluations.

Table K.1: Composition of the experiments with cyclohexane + toluene + acetone + methanol in mol %

		series A		series B		series C	
	$\bar{x}$	x_l	x_r	x_l	x_r	x_l	x_r
cyclohexane	5	5.5	4.5	5.5	4.5	5.5	4.5
toluene	5	5.5	4.5	5.5	4.5	4.5	5.5
acetone	5	6	4	4	6	5	5
methanol	85	78	82	85	85	85	85

Table K.2: Densities and viscosities of the system toluene + cyclohexane + acetone + methanol at 25 °C

x_1	x_2	x_3	density ρ in $\mathrm{g\,cm^{-3}}$	viscosity η in mPa s	variance of η in %
0.0550	0.0550	0.0601	0.7922	0.528	0.33
0.0450	0.0450	0.0400	0.7911	0.542	0.98
0.0450	0.0450	0.0601	0.7914	0.517	0.04
0.0550	0.0550	0.0400	0.7919	0.542	0.37
0.0450	0.0550	0.0501	0.7904	0.537	1.35
0.0550	0.0450	0.0501	0.7887	0.537	0.94
0.0520	0.0520	0.0740	0.7920	0.532	1.00
0.0475	0.0475	0.0351	0.7913	0.546	0.44
0.0475	0.0475	0.0600	0.7916	0.528	0.72
0.0525	0.0525	0.0400	0.7917	0.545	0.33
0.0475	0.0525	0.0501	0.7910	0.537	0.71
0.2198	0.1987	0.2095	0.7923	0.539	0.17

Table K.3: Densities and viscosities of the system toluene (1) + cyclohexane (2) + acetone (3) + methanol at x_1 =5.24 mol %, x_2 =5.24 mol %, and x_3 =5.2 mol % from 20 °C to 30 °C

T in °C	density ρ in $\mathrm{g\,cm^{-3}}$	viscosity η in mPa s	variance of η in %
20	0.7965	0.5740	3.68
24	0.79267	0.535	0.11
25	0.79167	0.529	0.05
26	0.79069	0.522	0.07
30	0.7868	0.5090	0.52

Table K.4: Partial molar volumes $\bar{V}_i$ of the system toluene (1) + cyclohexane (2) + acetone (3) + methanol at $x_1 = x_2 = x_3$ =5 mol % and molar volumes of the pure components V_i^0

	$\bar{V}_i$ in $cm^3\,mol^{-1}$	V_i^0
cyclohexane	112.3	108.8
toluene	112.9	106.9
acetone	71.91	74.05
methanol	40.82	40.74

Table K.5: Measured diffusion coefficient matrices for cyclohexane (1) + toluene (2) + acetone (3) + methanol at $x_1 = x_2 = x_3 = 5$ mol % and 298.15 K estimated from eqn (3.10). The quality of fit is given as $RMSE_{fit}$.

exp	D_{11}	D_{12}	D_{13}	D_{21}	D_{22} in $1 \times 10^{-9}\,m^2\,s^{-1}$	D_{23}	D_{31}	D_{32}	D_{33}	$RMSE_{fit}$ in mol %
1	6.2	-6.2	0.8	-29.4	41.7	-5.1	-105.2	144.1	-15.7	0.059
2	4.3	-3.1	0.1	5.5	-4.9	0.3	63.8	-75.6	5.4	0.039
3	-0.3	2.5	0.3	-10.2	17.4	3.5	31.1	-49.2	-9.7	0.042
4	7.2	-10.6	-2.9	4.0	-6.1	-2.5	-8.3	15.5	7.4	0.044
5	2.2	-0.9	-0.2	-1.4	4.5	0.3	1.4	-3.0	2.5	0.038
6	0.5	2.2	-1.3	-2.5	6.4	-1.3	-3.7	6.4	1.5	0.091
7	-1.8	4.0	-0.2	-5.8	9.7	-1.2	-2.0	2.8	2.5	0.037
8	-23.8	36.6	-6.9	-19.9	30.3	-5.1	-13.1	18.1	0.0	0.038
9	1.6	-0.6	2.0	-3.1	-0.4	0.2	-4.3	-3.6	5.5	0.047
10	1.7	-0.5	-0.8	0.0	2.6	0.0	-0.5	-0.8	2.8	0.040
11	145.8	143.6	-51.0	55.5	57.9	-15.7	202.5	202.7	-65.9	0.055
12	15.5	-19.3	-3.1	-35.0	52.2	8.7	-16.8	23.5	7.5	0.115
13	7.9	-10.2	-2.5	0.1	1.3	-0.5	-0.4	-0.6	2.9	0.037
14	4.8	0.5	2.1	-6.6	-0.6	-3.7	10.2	5.7	6.8	0.079
15	-164.2	-72.2	-398.8	62.9	29.7	152.2	59.1	25.6	143.6	0.042
16	-1.7	4.5	-30.3	1.1	1.4	9.0	0.9	-1.0	9.5	0.039

Table K.5: Measured diffusion coefficient matrices for cyclohexane (1) + toluene (2) + acetone (3) + methanol, continued.

exp	D_{11}	D_{12}	D_{13}	D_{21}	D_{22}	D_{23}	D_{31}	D_{32}	D_{33}	$RMSE_{fit}$
					in $1 \times 10^{-9}\,m^2\,s^{-1}$					in mol %
6 9	1.95	-0.30	-0.15	-0.17	2.27	-0.28	0.32	0.21	2.53	0.044
6 10	1.99	-0.36	-0.09	-0.20	2.37	-0.33	0.30	0.22	2.56	0.039
6 11	1.11	-1.36	1.44	-0.61	1.71	0.66	0.28	-0.26	2.70	0.054
6 14	1.93	-0.33	0.32	-0.48	3.20	-0.71	0.80	0.90	1.67	0.065
6 15	1.87	-0.21	-0.05	-0.38	2.36	-0.18	0.63	0.26	2.27	0.040
6 16	1.85	-0.45	0.13	-0.33	2.55	-0.35	0.63	0.32	2.21	0.038
6 3	2.24	-0.60	-0.16	-0.35	2.31	-0.25	1.86	-2.44	3.26	0.042
6 4	1.91	-0.34	-0.07	-0.18	2.23	-0.23	0.41	-0.75	3.16	0.041
6 5	1.22	0.60	-0.21	0.02	2.07	-0.27	1.98	-2.48	3.11	0.039
6 12	4.93	-2.89	-0.37	1.42	0.56	-0.31	-2.96	2.70	3.30	0.087
6 13	2.32	-0.26	-0.16	-0.95	2.95	-0.17	-1.98	1.70	3.22	0.043
7 9	1.80	-0.45	0.12	-0.10	2.34	-0.37	0.05	-0.03	2.93	0.042
7 10	1.84	-0.45	0.09	-0.12	2.47	-0.44	-0.04	-0.19	3.12	0.038
7 11	58.02	46.20	-58.85	27.02	24.73	-28.60	51.77	42.96	-52.35	0.058
7 14	1.94	-0.79	0.18	-0.47	2.42	-0.14	0.80	-0.41	2.69	0.062
7 15	1.80	-0.41	0.07	-0.39	2.37	-0.18	0.47	-0.07	2.64	0.038
7 16	1.80	-0.41	0.08	-0.39	2.41	-0.20	0.44	-0.11	2.70	0.036
7 3	1.48	0.07	-0.09	-0.05	2.17	-0.32	-0.02	-0.60	3.39	0.041
7 4	1.54	0.08	-0.06	-0.27	2.33	-0.23	0.19	-0.53	3.16	0.040
7 5	1.42	0.34	-0.23	0.41	1.66	-0.27	0.51	-0.95	3.21	0.037
7 12	2.20	-0.50	-0.14	0.08	1.92	-0.22	-2.37	1.97	3.29	0.086
7 13	2.72	-0.90	-0.01	-0.33	2.36	-0.21	-0.08	-0.20	3.14	0.037
8 9	2.18	-0.15	-0.51	-0.33	2.14	-0.05	-0.16	-0.19	3.10	0.043
8 10	2.33	0.10	-0.78	-0.17	2.40	-0.36	-0.25	-0.37	3.29	0.038
8 11	119.40	117.01	-133.92	84.60	87.43	-96.96	142.90	143.21	-161.31	0.069
8 14	1.97	-0.28	-0.12	-0.51	2.73	-0.35	0.87	-0.95	2.87	0.062
8 15	2.06	0.09	-0.58	-0.29	2.52	-0.36	0.26	-0.40	2.93	0.039
8 16	2.06	0.19	-0.64	-0.35	2.49	-0.29	0.22	-0.46	3.00	0.037
8 3	1.44	0.07	-0.07	-1.10	3.24	-0.31	3.31	-4.02	3.34	0.040
8 4	3.11	-1.50	-0.31	-0.13	2.19	-0.23	-0.13	-0.27	3.13	0.041

Table K.5: Measured diffusion coefficient matrices for cyclohexane (1) + toluene (2) + acetone (3) + methanol, continued.

exp	D_{11}	D_{12}	D_{13}	D_{21}	D_{22}	D_{23}	D_{31}	D_{32}	D_{33}	$RMSE_{fit}$
					in $1 \times 10^{-9}\,m^2\,s^{-1}$					in mol %
8 5	1.92	-0.06	-0.32	0.23	1.80	-0.24	-0.81	0.20	3.24	0.039
8 12	5.71	-4.08	-0.27	0.66	1.27	-0.20	-2.09	1.50	3.28	0.086
8 13	3.09	-1.71	-0.05	-0.39	2.34	-0.21	1.34	-1.65	3.08	0.037
1 9	1.94	-0.32	0.04	-0.21	2.25	-0.13	0.81	0.64	2.22	0.053
1 10	1.96	-0.34	0.03	-0.24	2.31	-0.15	0.72	0.71	2.24	0.049
1 11	7.54	4.61	-6.83	11.09	13.05	-14.20	7.74	7.35	-7.51	0.066
1 14	2.00	-0.67	0.25	-0.40	2.27	0.03	1.00	0.87	1.86	0.070
1 15	1.92	-0.26	0.00	-0.38	2.33	-0.04	0.98	0.68	2.03	0.050
1 16	1.90	-0.25	0.01	-0.39	2.36	-0.06	0.92	0.64	2.12	0.048
1 3	0.22	1.54	0.07	-1.99	4.30	-0.12	4.97	-5.53	3.33	0.051
1 4	1.84	-0.14	0.00	-0.19	2.29	-0.20	-0.13	0.16	3.40	0.052
1 5	1.07	0.83	-0.18	-0.94	3.13	-0.17	5.19	-5.37	3.06	0.050
1 12	2.39	-0.70	-0.09	-0.47	2.55	-0.15	-0.67	0.58	3.52	0.092
1 13	3.05	-1.54	-0.02	-0.39	2.48	-0.13	0.95	-0.93	3.30	0.049
2 9	1.99	-0.31	-0.09	-0.28	2.20	-0.04	-0.03	-0.11	3.04	0.043
2 10	2.01	-0.29	-0.06	-0.27	2.27	-0.11	-0.11	-0.21	3.18	0.038
2 11	108.75	80.72	-106.64	22.39	19.00	-22.16	82.88	62.97	-80.65	0.053
2 14	1.96	-0.63	0.19	-0.41	2.83	-0.37	0.79	-1.07	3.09	0.062
2 15	1.96	-0.19	-0.08	-0.40	2.28	-0.02	0.42	-0.17	2.76	0.039
2 16	1.97	-0.14	-0.10	-0.42	2.29	-0.01	0.37	-0.25	2.85	0.037
2 3	0.81	0.94	0.11	-1.44	3.58	-0.21	1.82	-2.49	3.29	0.040
2 4	1.72	-0.05	-0.01	-0.48	2.62	-0.18	-0.06	-0.27	3.18	0.041
2 5	1.40	0.43	-0.14	-0.09	2.25	-0.22	0.57	-1.04	3.21	0.038
2 12	2.23	-0.53	-0.10	-0.37	2.45	-0.16	-1.00	0.57	3.29	0.086
2 13	2.22	-0.59	0.01	-0.31	2.39	-0.15	-0.10	-0.15	3.12	0.037
3 9	1.59	-0.68	-0.50	-0.69	1.73	-1.11	0.58	0.61	4.69	0.044
3 10	1.64	-0.63	-0.32	-0.69	1.93	-0.98	0.69	0.45	4.66	0.040
3 11	37.18	31.29	49.81	9.67	10.99	13.50	85.85	76.91	125.67	0.046
3 14	1.56	-0.43	-0.33	-0.77	2.54	-0.56	1.16	-1.33	3.64	0.065
3 15	1.35	-0.51	-0.54	-0.62	2.30	-0.62	1.04	-0.15	4.45	0.041

Table K.5: Measured diffusion coefficient matrices for cyclohexane (1) + toluene (2) + acetone (3) + methanol, continued.

exp	D_{11}	D_{12}	D_{13}	D_{21}	D_{22}	D_{23}	D_{31}	D_{32}	D_{33}	$RMSE_{fit}$
				in $1 \times 10^{-9}\,m^2\,s^{-1}$						in mol %
3 16	1.34	-0.49	-0.56	-0.71	2.28	-0.70	1.12	-0.13	4.53	0.038
4 9	1.93	-0.34	-0.07	-0.17	2.28	-0.19	-0.15	-0.17	3.17	0.045
4 10	1.94	-0.35	-0.08	-0.24	2.31	-0.22	-0.09	-0.19	3.21	0.040
4 11	2.95	0.60	0.93	1.62	3.78	2.05	6.05	4.99	10.96	0.056
4 14	2.03	-0.39	0.12	-0.43	2.54	-0.21	0.21	-1.23	2.68	0.064
4 15	1.89	-0.28	-0.06	-0.32	2.43	-0.20	0.15	-0.58	3.10	0.041
4 16	1.87	-0.27	-0.07	-0.34	2.46	-0.19	0.17	-0.56	3.13	0.039
5 9	1.88	-0.42	-0.51	-0.16	2.28	-0.27	0.26	0.25	3.88	0.043
5 10	1.90	-0.34	-0.27	-0.14	2.43	-0.14	0.06	-0.16	3.42	0.038
5 11	2.14	-0.39	-1.17	0.94	3.03	1.45	0.66	0.59	4.34	0.072
5 14	1.75	-0.25	-0.28	-0.20	2.62	-0.06	1.06	-1.17	3.45	0.062
5 15	1.77	-0.31	-0.32	-0.02	2.65	0.10	0.53	-0.37	3.65	0.038
5 16	1.72	-0.25	-0.35	-0.04	2.68	0.10	0.62	-0.33	3.74	0.036
12 9	2.14	-0.17	0.23	-0.29	2.16	-0.28	-0.78	-0.81	2.44	0.088
12 10	2.13	-0.19	0.22	-0.42	2.15	-0.40	-0.52	-0.58	2.81	0.086
12 11	1.76	-0.57	-0.67	1.75	4.28	2.81	0.46	0.13	3.86	0.095
12 14	2.24	-0.51	0.16	-0.48	2.47	-0.21	-0.59	-1.44	2.11	0.099
12 15	2.07	-0.20	0.14	-0.48	2.30	-0.31	-0.57	-1.08	2.39	0.086
12 16	2.04	-0.19	0.12	-0.49	2.34	-0.29	-0.56	-1.06	2.43	0.085
13 9	2.13	-0.14	0.30	-0.19	2.26	-0.15	-0.28	-0.31	2.85	0.042
13 10	2.19	-0.12	0.34	-0.28	2.27	-0.22	-0.15	-0.23	3.02	0.036
13 11	22.92	22.06	31.31	18.90	22.76	28.10	34.19	36.50	53.83	0.044
13 14	2.27	-0.50	0.18	-0.55	2.51	-0.28	-0.36	-1.28	2.16	0.062
13 15	2.05	-0.23	0.16	-0.25	2.45	-0.07	-0.18	-0.75	2.63	0.037
13 16	2.01	-0.25	0.13	-0.27	2.47	-0.07	-0.16	-0.74	2.66	0.035
6 3 10	1.94	-0.35	-0.03	-0.22	2.34	-0.29	-0.19	-0.32	3.37	0.040
6 3 11	1.96	-0.35	-0.05	-0.22	2.33	-0.29	-0.24	-0.27	3.39	0.050
6 3 14	1.92	-0.35	-0.04	-0.44	2.59	-0.27	0.79	-1.37	3.32	0.058
6 3 15	1.88	-0.29	-0.04	-0.32	2.46	-0.29	0.15	-0.68	3.37	0.041
6 3 16	1.87	-0.28	-0.04	-0.35	2.49	-0.28	0.16	-0.69	3.37	0.040

Table K.5: Measured diffusion coefficient matrices for cyclohexane (1) + toluene (2) + acetone (3) + methanol, continued.

exp	D_{11}	D_{12}	D_{13}	D_{21}	D_{22}	D_{23}	D_{31}	D_{32}	D_{33}	$RMSE_{fit}$
					in $1 \times 10^{-9}\,m^2\,s^{-1}$					in mol %
6 3 9	1.93	-0.33	-0.04	-0.16	2.28	-0.30	-0.25	-0.27	3.39	0.044
6 4 10	1.96	-0.33	-0.06	-0.25	2.30	-0.24	-0.09	-0.20	3.19	0.040
6 4 11	1.97	-0.02	-0.03	0.49	1.77	-0.25	-0.18	-0.10	3.19	0.056
6 4 14	1.93	-0.32	-0.06	-0.44	2.51	-0.22	0.78	-1.13	3.14	0.057
6 4 15	1.90	-0.27	-0.06	-0.35	2.41	-0.23	0.22	-0.53	3.18	0.041
6 4 16	1.88	-0.26	-0.06	-0.37	2.44	-0.23	0.22	-0.53	3.18	0.039
6 4 9	1.95	-0.32	-0.06	-0.20	2.25	-0.24	-0.13	-0.16	3.19	0.043
6 5 10	2.01	-0.28	-0.14	-0.22	2.34	-0.28	-0.13	-0.24	3.27	0.038
6 5 11	1.85	0.02	-0.15	0.63	1.76	-0.16	-0.08	-0.28	3.27	0.056
6 5 14	1.93	-0.21	-0.14	-0.44	2.59	-0.27	0.78	-1.23	3.21	0.056
6 5 15	1.93	-0.20	-0.14	-0.33	2.45	-0.27	0.20	-0.58	3.25	0.039
6 5 16	1.92	-0.19	-0.14	-0.35	2.48	-0.27	0.20	-0.59	3.25	0.038
6 5 9	2.00	-0.28	-0.14	-0.16	2.27	-0.28	-0.17	-0.20	3.27	0.042
6 12 10	1.96	-0.34	-0.05	-0.28	2.27	-0.18	-0.16	-0.27	3.29	0.074
6 12 11	1.88	-0.02	0.04	0.52	1.75	-0.05	-0.11	-0.40	3.25	0.084
6 12 14	1.94	-0.32	-0.06	-0.45	2.45	-0.17	0.77	-1.24	3.22	0.085
6 12 15	1.90	-0.27	-0.06	-0.37	2.37	-0.17	0.17	-0.60	3.26	0.074
6 12 16	1.88	-0.26	-0.05	-0.39	2.39	-0.17	0.17	-0.61	3.26	0.073
6 12 9	1.95	-0.33	-0.05	-0.23	2.22	-0.19	-0.21	-0.22	3.29	0.076
6 13 10	1.93	-0.36	-0.01	-0.27	2.28	-0.20	-0.05	-0.15	3.12	0.037
6 13 11	1.94	-0.04	-0.07	0.62	1.64	-0.02	0.35	0.15	2.94	0.068
6 13 14	1.94	-0.38	-0.01	-0.45	2.47	-0.18	0.77	-1.05	3.08	0.055
6 13 15	1.88	-0.31	-0.02	-0.36	2.38	-0.19	0.25	-0.48	3.11	0.038
6 13 16	1.87	-0.29	-0.02	-0.39	2.41	-0.19	0.25	-0.48	3.11	0.037
6 13 9	1.92	-0.35	-0.02	-0.22	2.23	-0.20	-0.08	-0.12	3.13	0.041
7 3 10	1.93	-0.36	-0.05	-0.22	2.34	-0.29	-0.22	-0.34	3.34	0.039
7 3 11	1.55	-0.08	-0.04	0.01	2.15	-0.35	-0.10	-0.47	3.37	0.050
7 3 14	1.91	-0.35	-0.04	-0.44	2.58	-0.26	0.76	-1.36	3.29	0.056
7 3 15	1.87	-0.30	-0.05	-0.32	2.46	-0.29	0.13	-0.70	3.34	0.040
7 3 16	1.86	-0.28	-0.05	-0.35	2.49	-0.29	0.13	-0.71	3.34	0.039

Table K.5: Measured diffusion coefficient matrices for cyclohexane (1) + toluene (2) + acetone (3) + methanol, continued.

exp	D_{11}	D_{12}	D_{13}	D_{21}	D_{22}	D_{23}	D_{31}	D_{32}	D_{33}	$RMSE_{fit}$
					in $1 \times 10^{-9}\,m^2\,s^{-1}$					in mol %
7 3 9	1.92	-0.34	-0.05	-0.16	2.28	-0.29	-0.27	-0.29	3.35	0.043
7 4 10	1.95	-0.34	-0.07	-0.25	2.30	-0.24	-0.12	-0.22	3.18	0.039
7 4 11	1.83	0.00	-0.02	0.20	1.93	-0.19	-0.10	-0.24	3.17	0.056
7 4 14	1.93	-0.32	-0.06	-0.45	2.52	-0.22	0.77	-1.14	3.13	0.056
7 4 15	1.89	-0.28	-0.07	-0.35	2.41	-0.23	0.21	-0.55	3.16	0.040
7 4 16	1.87	-0.27	-0.07	-0.37	2.44	-0.23	0.20	-0.54	3.16	0.038
7 4 9	1.94	-0.33	-0.08	-0.19	2.25	-0.24	-0.15	-0.18	3.18	0.042
7 5 10	2.00	-0.28	-0.15	-0.22	2.33	-0.28	-0.15	-0.26	3.24	0.037
7 5 11	1.89	0.01	-0.08	0.49	1.83	-0.23	-0.11	-0.28	3.25	0.053
7 5 14	1.93	-0.21	-0.14	-0.45	2.58	-0.27	0.77	-1.23	3.20	0.055
7 5 15	1.93	-0.21	-0.15	-0.33	2.45	-0.27	0.18	-0.60	3.23	0.038
7 5 16	1.91	-0.19	-0.15	-0.35	2.48	-0.27	0.18	-0.60	3.23	0.036
7 5 9	2.00	-0.28	-0.14	-0.16	2.27	-0.28	-0.19	-0.22	3.24	0.041
7 12 10	1.95	-0.35	-0.06	-0.28	2.27	-0.19	-0.18	-0.29	3.26	0.073
7 12 11	2.14	-0.05	-0.18	0.55	1.59	0.12	-0.04	-0.43	3.48	0.089
7 12 14	1.94	-0.34	-0.06	-0.46	2.45	-0.18	0.76	-1.25	3.21	0.084
7 12 15	1.89	-0.28	-0.06	-0.37	2.36	-0.18	0.15	-0.62	3.24	0.074
7 12 16	1.87	-0.27	-0.06	-0.39	2.39	-0.18	0.15	-0.62	3.24	0.073
7 12 9	1.94	-0.34	-0.06	-0.23	2.22	-0.19	-0.23	-0.24	3.26	0.075
7 13 10	1.93	-0.37	-0.03	-0.27	2.28	-0.20	-0.08	-0.16	3.11	0.036
7 13 11	1.93	-0.05	0.08	0.39	1.72	-0.11	0.02	-0.26	3.13	0.053
7 13 14	1.93	-0.37	-0.02	-0.46	2.47	-0.19	0.76	-1.06	3.08	0.054
7 13 15	1.87	-0.31	-0.03	-0.37	2.38	-0.20	0.23	-0.49	3.09	0.037
7 13 16	1.86	-0.30	-0.03	-0.39	2.41	-0.20	0.23	-0.49	3.10	0.036
7 13 9	1.91	-0.36	-0.03	-0.22	2.23	-0.21	-0.10	-0.14	3.12	0.040
8 3 10	1.89	-0.40	-0.11	-0.22	2.35	-0.28	-0.25	-0.38	3.31	0.040
8 3 11	1.67	-0.17	0.04	-0.10	2.27	-0.41	0.01	-0.69	3.41	0.051
8 3 14	1.85	-0.35	-0.10	-0.44	2.58	-0.27	0.73	-1.37	3.27	0.057
8 3 15	1.83	-0.33	-0.11	-0.32	2.46	-0.29	0.09	-0.72	3.31	0.040
8 3 16	1.81	-0.31	-0.11	-0.35	2.49	-0.29	0.10	-0.73	3.31	0.039

Table K.5: Measured diffusion coefficient matrices for cyclohexane (1) + toluene (2) + acetone (3) + methanol, continued.

exp	D_{11}	D_{12}	D_{13}	D_{21}	D_{22}	D_{23}	D_{31}	D_{32}	D_{33}	$\mathrm{RMSE_{fit}}$
				in $1 \times 10^{-9}\,\mathrm{m^2\,s^{-1}}$						in mol %
8 3 9	1.89	-0.39	-0.11	-0.16	2.28	-0.29	-0.30	-0.33	3.32	0.043
8 4 10	1.91	-0.38	-0.13	-0.24	2.31	-0.23	-0.15	-0.24	3.13	0.039
8 4 11	1.77	0.02	-0.10	0.65	1.68	-0.09	-0.09	-0.33	3.13	0.057
8 4 14	1.87	-0.33	-0.11	-0.46	2.52	-0.23	0.73	-1.15	3.11	0.056
8 4 15	1.84	-0.31	-0.13	-0.35	2.41	-0.23	0.17	-0.57	3.12	0.040
8 4 16	1.83	-0.29	-0.13	-0.37	2.44	-0.23	0.17	-0.57	3.12	0.039
8 4 9	1.90	-0.38	-0.13	-0.19	2.25	-0.23	-0.19	-0.21	3.13	0.043
8 5 10	1.96	-0.32	-0.21	-0.22	2.34	-0.28	-0.19	-0.29	3.20	0.038
8 5 11	2.12	-0.13	0.14	0.70	1.61	-0.11	0.01	-0.53	3.16	0.063
8 5 14	1.87	-0.22	-0.20	-0.46	2.59	-0.28	0.74	-1.23	3.18	0.055
8 5 15	1.88	-0.24	-0.21	-0.33	2.45	-0.27	0.14	-0.63	3.19	0.038
8 5 16	1.86	-0.22	-0.21	-0.35	2.48	-0.27	0.14	-0.63	3.19	0.037
8 5 9	1.96	-0.33	-0.19	-0.16	2.28	-0.27	-0.23	-0.25	3.19	0.041
8 12 10	1.91	-0.38	-0.13	-0.28	2.27	-0.18	-0.23	-0.32	3.22	0.073
8 12 11	2.00	0.02	-0.21	0.56	1.64	0.10	-0.17	-0.45	3.33	0.088
8 12 14	1.88	-0.35	-0.11	-0.47	2.46	-0.20	0.73	-1.26	3.20	0.084
8 12 15	1.85	-0.31	-0.13	-0.37	2.36	-0.18	0.11	-0.65	3.21	0.074
8 12 16	1.83	-0.29	-0.13	-0.40	2.39	-0.18	0.11	-0.65	3.21	0.073
8 12 9	1.91	-0.37	-0.12	-0.22	2.22	-0.19	-0.27	-0.28	3.22	0.075
8 13 10	1.88	-0.41	-0.08	-0.27	2.28	-0.20	-0.11	-0.19	3.06	0.037
8 13 11	-452.41	429.67	193.93	-372.91	354.65	158.84	-240.70	227.48	105.42	0.146
8 13 14	1.87	-0.38	-0.08	-0.46	2.47	-0.19	0.73	-1.06	3.05	0.055
8 13 15	1.83	-0.35	-0.09	-0.37	2.38	-0.20	0.19	-0.51	3.06	0.038
8 13 16	1.81	-0.33	-0.09	-0.39	2.40	-0.20	0.19	-0.51	3.06	0.036
8 13 9	1.88	-0.40	-0.09	-0.22	2.23	-0.20	-0.14	-0.17	3.06	0.040
1 3 10	1.97	-0.32	0.01	-0.19	2.37	-0.25	-0.03	-0.18	3.60	0.048
1 3 11	1.98	-0.18	-0.01	0.03	2.16	-0.22	-0.03	-0.24	3.60	0.056
1 3 14	1.96	-0.35	0.00	-0.38	2.60	-0.23	1.05	-1.34	3.55	0.062
1 3 15	1.91	-0.28	0.01	-0.29	2.49	-0.24	0.33	-0.57	3.59	0.048
1 3 16	1.90	-0.26	0.01	-0.31	2.51	-0.24	0.34	-0.58	3.59	0.047

Table K.5: Measured diffusion coefficient matrices for cyclohexane (1) + toluene (2) + acetone (3) + methanol, continued.

exp	D_{11}	D_{12}	D_{13}	D_{21}	D_{22}	D_{23}	D_{31}	D_{32}	D_{33}	$\mathrm{RMSE_{fit}}$
					in $1 \times 10^{-9}\,\mathrm{m^2\,s^{-1}}$					in mol %
1 3 9	1.96	-0.31	0.01	-0.15	2.32	-0.25	-0.10	-0.11	3.60	0.050
1 4 10	1.98	-0.32	-0.03	-0.22	2.34	-0.19	0.07	-0.05	3.39	0.047
1 4 11	-210.75	299.69	56.73	-145.93	207.74	38.72	-22.56	30.90	9.07	0.138
1 4 14	1.98	-0.32	-0.03	-0.39	2.53	-0.17	1.03	-1.09	3.34	0.063
1 4 15	1.93	-0.25	-0.02	-0.31	2.44	-0.18	0.40	-0.40	3.37	0.048
1 4 16	1.91	-0.24	-0.02	-0.33	2.46	-0.18	0.41	-0.41	3.37	0.047
1 4 9	1.97	-0.30	-0.01	-0.17	2.28	-0.19	0.03	0.00	3.39	0.050
1 5 10	2.03	-0.26	-0.11	-0.19	2.36	-0.24	0.03	-0.10	3.47	0.046
1 5 11	2.03	-0.12	-0.08	-0.01	2.17	-0.19	0.20	-0.27	3.52	0.055
1 5 14	1.97	-0.22	-0.12	-0.39	2.60	-0.22	1.04	-1.18	3.42	0.061
1 5 15	1.96	-0.18	-0.11	-0.29	2.48	-0.22	0.38	-0.47	3.45	0.046
1 5 16	1.95	-0.17	-0.10	-0.31	2.50	-0.22	0.38	-0.46	3.45	0.045
1 5 9	2.01	-0.26	-0.11	-0.16	2.32	-0.23	-0.01	-0.07	3.44	0.049
1 12 10	1.99	-0.32	-0.04	-0.25	2.31	-0.14	-0.04	-0.12	3.50	0.078
1 12 11	-257.24	428.21	84.61	-133.97	223.46	43.51	-113.29	186.24	40.10	0.159
1 12 14	1.97	-0.30	-0.04	-0.37	2.43	-0.10	1.01	-1.18	3.41	0.088
1 12 15	1.93	-0.25	-0.02	-0.33	2.39	-0.12	0.35	-0.49	3.46	0.078
1 12 16	1.92	-0.24	-0.02	-0.35	2.41	-0.12	0.35	-0.49	3.47	0.078
1 12 9	1.97	-0.30	-0.01	-0.20	2.25	-0.14	-0.06	-0.07	3.49	0.080
1 13 10	1.96	-0.34	0.03	-0.24	2.31	-0.15	0.10	0.00	3.32	0.045
1 13 11	0.65	3.95	1.46	0.39	2.99	0.62	-0.39	0.53	3.54	0.128
1 13 14	1.98	-0.36	0.01	-0.38	2.47	-0.13	1.02	-1.01	3.28	0.061
1 13 15	1.90	-0.25	-0.02	-0.34	2.42	-0.16	0.42	-0.34	3.34	0.046
1 13 16	1.89	-0.24	-0.02	-0.36	2.44	-0.16	0.42	-0.34	3.34	0.045
1 13 9	1.95	-0.32	0.03	-0.19	2.26	-0.15	0.07	0.04	3.32	0.048
2 3 10	1.96	-0.33	-0.01	-0.19	2.37	-0.24	-0.22	-0.35	3.36	0.040
2 3 11	1.95	-0.14	0.00	0.01	2.00	-0.31	-0.02	-0.62	3.43	0.052
2 3 14	1.96	-0.34	0.00	-0.38	2.59	-0.22	0.74	-1.37	3.28	0.057
2 3 15	1.91	-0.28	0.00	-0.28	2.49	-0.24	0.12	-0.71	3.35	0.041
2 3 16	1.89	-0.26	0.00	-0.31	2.51	-0.24	0.12	-0.72	3.35	0.039

Table K.5: Measured diffusion coefficient matrices for cyclohexane (1) + toluene (2) + acetone (3) + methanol, continued.

exp	D_{11}	D_{12}	D_{13}	D_{21}	D_{22}	D_{23}	D_{31}	D_{32}	D_{33}	RMSE_{fit}
				in $1 \times 10^{-9}\,\text{m}^2\,\text{s}^{-1}$						in mol %
2 3 9	1.95	-0.32	-0.01	-0.13	2.31	-0.25	-0.28	-0.29	3.37	0.043
2 4 10	1.98	-0.31	-0.03	-0.21	2.33	-0.19	-0.11	-0.21	3.18	0.040
2 4 11	-414.78	453.45	145.11	-306.37	335.50	106.43	-243.75	264.67	87.63	0.137
2 4 14	1.98	-0.31	-0.02	-0.40	2.53	-0.18	0.76	-1.14	3.13	0.057
2 4 15	1.93	-0.26	-0.02	-0.31	2.43	-0.18	0.20	-0.55	3.16	0.040
2 4 16	1.91	-0.24	-0.02	-0.33	2.46	-0.18	0.20	-0.55	3.16	0.039
2 4 9	1.97	-0.30	-0.03	-0.16	2.28	-0.19	-0.15	-0.17	3.18	0.043
2 5 10	2.03	-0.25	-0.11	-0.19	2.37	-0.23	-0.15	-0.26	3.25	0.038
2 5 11	-445.25	484.30	164.66	-314.18	342.27	115.38	-295.15	319.44	111.51	0.145
2 5 14	1.99	-0.20	-0.09	-0.40	2.60	-0.23	0.76	-1.23	3.19	0.055
2 5 15	1.97	-0.18	-0.10	-0.29	2.47	-0.23	0.18	-0.61	3.23	0.039
2 5 16	1.95	-0.17	-0.10	-0.31	2.50	-0.23	0.17	-0.61	3.23	0.037
2 5 9	2.03	-0.25	-0.10	-0.13	2.31	-0.23	-0.19	-0.21	3.25	0.041
2 12 10	1.98	-0.32	-0.02	-0.25	2.30	-0.15	-0.19	-0.29	3.26	0.074
2 12 11	-29.33	63.02	4.09	-14.67	31.66	1.87	-20.00	39.20	6.55	0.161
2 12 14	2.00	-0.32	-0.01	-0.42	2.46	-0.15	0.75	-1.26	3.21	0.084
2 12 15	1.93	-0.26	-0.02	-0.34	2.39	-0.13	0.14	-0.62	3.24	0.074
2 12 16	1.91	-0.24	-0.02	-0.36	2.41	-0.13	0.15	-0.63	3.24	0.073
2 12 9	1.97	-0.31	-0.02	-0.20	2.25	-0.15	-0.23	-0.24	3.27	0.075
2 13 10	1.96	-0.34	0.02	-0.24	2.31	-0.16	-0.07	-0.16	3.11	0.037
2 13 11	-370.32	470.83	154.83	-225.88	287.82	93.72	-208.77	263.45	89.48	0.143
2 13 14	1.98	-0.37	0.02	-0.40	2.48	-0.14	0.76	-1.06	3.07	0.055
2 13 15	1.91	-0.27	0.00	-0.33	2.42	-0.16	0.22	-0.52	3.12	0.038
2 13 16	1.90	-0.26	0.00	-0.35	2.44	-0.16	0.22	-0.52	3.12	0.036
2 13 9	1.95	-0.33	0.01	-0.20	2.27	-0.16	-0.10	-0.15	3.12	0.041

Table K.6: Measured diffusion coefficient matrices for cyclohexane (1) + toluene (2) + acetone (3) + methanol at $x_1 = x_2 = x_3 = 5\,\mathrm{mol\%}$ and 298.15 K estimated from eqn (3.10) with standard deviations from repeated experiments. For the diffusion coefficients matrices, also the $\mathrm{RMSE_{fit}}$ is given.

	D_{11}	D_{12}	D_{13}	D_{21}	D_{22}	D_{23}	D_{31}	D_{32}	D_{33}	$\mathrm{RMSE_{fit}}$
					in $1 / 10^{-9}\,\mathrm{m^2\,s^{-1}}$					in mol %
$n_{exp} = 1$[a]	0.4(573)	4.4(429)	−30.8(992)	1.0(255)	15.2(207)	8.7(367)	13.4(624)	19.4(663)	6.7(408)	0.053
$n_{exp} = 2$[b]	2.076(932)	−0.308(894)	−0.163(827)	−0.126(1373)	2.557(1297)	−0.3198(16520)	0.504(1574)	−0.255(1643)	3.027(1598)	0.051
$n_{exp} = 3$[c]	1.923(68)	−0.272(95)	−0.060(61)	−0.231(246)	2.335(211)	−0.209(69)	0.165(351)	−0.523(368)	3.260(132)	0.052
$n_{exp} = 16$[d]	1.810	−0.033	−0.012	0.464	2.142	0.138	0.048	−0.470	3.084	0.062
$n_{exp} = 12$[e]	1.615	0.664	1.298	−0.153	3.146	1.029	−0.026	0.026	3.250	0.128
$n_{exp} = 12$[f]	1.884	−0.243	−0.046	0.081	2.170	0.031	0.187	−0.584	3.174	0.047

[a] estimated from 16 separate experiments.
[b] estimated from 82 combinations of two independent experiments; 87 evaluations were available.
[c] estimated from 142 combinations of three independent experiments; 150 evaluations were available.
[d] estimated from a total of 16 experiments using series A, B, and C.
[e] estimated from a total of 12 experiments using series A, B, and C.
Experiments 6, 10, 12, and 14 were excluded to increase the quality of fit,
but a disadvantageous starting value was chosen for $\boldsymbol{D}^V$. Thus a large $\mathrm{RMSE_{fit}}$.
[f] estimated from a total of 12 experiments using series A, B, and C.
Experiments 6, 10, 12, and 14 were excluded to increase the quality of fit.
For *d*, *e* and *f*, no standard deviations from repeated evaluations are available.
For *d* and *e* standard deviations are expected to be larger as the $\mathrm{RMSE_{fit}}$ is larger.
But for *f*, standard deviations are expected to be similar to $n_{exp} = 3$ as more data was used and the $\mathrm{RMSE_{fit}}$ is similar.

Bibliography

[1] T. Großmann and J. Winkelmann, "Ternary diffusion coefficients of cyclohexane + toluene + methanol by Taylor dispersion measurements at 298.15 K. part 2. low toluene area near the binodal curve," J. Chem. Eng. Data, vol. 54, pp. 485–490, 2009.

[2] S. Käshammer, H. Weingärtner, and H. G. Hertz, "Ternary diffusion in the system n-propanol + 1-chlorobutane + n-heptane at 25 °C," Z. Phys. Chem., vol. 187, pp. 233–255, 1994.

[3] R. Taylor and R. Krishna, Multicomponent Mass Transfer. JOHN WILEY & SONS, INC., 1993.

[4] M. Wellhausen, G. Rinke, and H. Wackerbarth, "Combined measurement of concentration distribution and velocity field of two components in a micromixing process," Microfluid. Nanofluid., vol. 12, pp. 917–926, 2012.

[5] Y. Wang, Q. Lin, and T. Mukherjee, "A model for laminar diffusion-based complex electrokinetic passive micromixers," Lab Chip, vol. 5, pp. 877–887, 2005.

[6] A. F. Chrimes, K. Khoshmanesh, P. R. Stoddart, A. Mitchella, and K. Kalantar-zadeh, "Microfluidics and Raman microscopy: current applications and future challenges," Chem. Soc. Rev., vol. 42, pp. 5880–5906, 2013.

[7] N. L. Jeon, S. K. Dertinger, D. T. Chiu, I. S. Choi, A. D. Stroock, and G. M. Whitesides, "Generation of solution and surface gradients using microfluidic systems," Langmuir, vol. 16, pp. 8311–8316, 2000.

[8] J. Yang, X. Pi, L. Zhang, X. Liu, J. Yang, Y. Cao, W. Zhang, and X. Zheng, "Diffusion characteristics of a T-type microchannel with different configurations and inlet angles," Anal. Sci., vol. 23, pp. 697–703, 2007.

[9] C. Kim, K. Kreppenhofer, J. Kashef, D. Gradl, D. Herrmann, M. Schneider, R. Ahrens, A. Guberb, and D. Wedlich, "Diffusion- and convection-based acti-

vation of Wnt/β-catenin signaling in a gradient generating microfluidic chip," Lab Chip, vol. 12, pp. 5186–5194, 2012.

[10] A. E. Kamholz, E. A. Schilling, and P. Yager, "Optical measurement of transverse molecular diffusion in a microchannel," Biophys. J., vol. 80, pp. 1967–1972, 2001.

[11] M. Yang, J. Yang, C.-W. Li, and J. Zhao, "Generation of concentration gradient by controlled flow distribution and diffusive mixing in a microfluidic chip," Lab Chip, vol. 2, pp. 158–163, 2002.

[12] T. Frank and S. Tay, "Flow-switching allows independently programmable, extremely stable, high-throughput diffusion-based gradients," Lab Chip, vol. 13, p. 1273, 2013.

[13] J. Atencia, J. Morrow, and L. E. Locascio, "The microfluidic palette: a diffusive gradient generator with spatio-temporal control.," Lab Chip, vol. 9, pp. 2707–14, Sept. 2009.

[14] J. L. Osborn, B. Lutz, E. Fu, P. Kauffman, D. Y. Stevens, and P. Yager, "Microfluidics without pumps: reinventing the T-sensor and H-filter in paper networks," Lab Chip, vol. 10, pp. 2659–2665, 2010.

[15] E. Berthier and D. J. Beebe, "Gradient generation platforms: new directions for an established microfluidic technology," Lab Chip, vol. 14, pp. 3241–3247, 2014.

[16] I. G. Economou, J.-C. de Hemptinne, R. Dohrn, E. Hendriks, K. Keskinen, and O. Baudouin, "Industrial use of thermodynamics workshop: Round table discussion on 8 July 2014," Chem. Eng. Res. Des., vol. 92, pp. 2795 – 2796, 2014. Advances in Thermodynamics for Chemical Process and Product Design.

[17] C. Peters, L. Wolff, S. Haase, J. Thien, T. Brands, H.-J. Koß, and A. Bardow, "Multicomponent diffusion coefficients from microfluidics using Raman microspectroscopy," Lab Chip, vol. 17, pp. 2768–2776, 2017.

[18] N. Miložič, M. Lubej, U. Novak, P. Žnidaršič-Plazl, and I. Plazl, "Evaluation of diffusion coefficient determination using a microfluidic device," Chem. Biochem. Eng. Q., vol. 28, pp. 215–223, 2014.

[19] A. A. Shapiro, "Evaluation of diffusion coefficients in multicomponent mixtures by means of the fluctuaction theory," Physica A, vol. 320, pp. 211–234, 2003.

[20] R. B. Bird, "Five decades of transport phenomena," AIChE J., vol. 50, pp. 273–287, Feb. 2004.

[21] DDBST Dortmund Data Bank Software & Separation Technology GmbH, "Statistics on online ddb search," 02.01.2017.

[22] J. Winkelmann, Diffusion in Gases, Liquids and Electrolytes; Nonelectrolyte Liquids and Liquid Mixtures - Part 2: Liquid Mixtures. Springer-Verlag Berlin Heidelberg, 2017.

[23] J. W. Mutoru and A. Firoozabadi, "Supplementary data for "form of multicomponent Fickian diffusion coefficients matrix"," J. Chem. Thermodyn., vol. 43, pp. 1192–1203, 2011.

[24] X. Liu, S. K. Schnell, J.-M. Simon, D. Bedeaux, S. Kjelstrup, A. Bardow, and T. J. H. Vlugt, "Fick diffusion coefficients of liquid mixtures directly obtained from equilibrium molecular dynamics," J. Phys. Chem. B, vol. 115, pp. 12921–12929, 2011.

[25] H. Song and D. Cabooter, "Relevance and assessment of molecular diffusion coefficients in liquid chromatography," Chromatographia, vol. 5, pp. 651–663, 2016.

[26] X. Liu, T. J. Vlugt, and A. Bardow, "Predictive Darken equation for Maxwell-Stefan diffusivities in multicomponent mixtures," Ind. Eng. Chem. Res., vol. 50, pp. 10350–10358, 2011.

[27] C. Peters, L. Wolff, T. J. H. Vlugt, and A. Bardow, Experimental Thermodynamics Volume X : Non-equilibrium Thermodynamics with Applications, ch. 5 - Diffusion in Liquids : Experiments, Molecular Dynamics, and Engineering Models, pp. 78–104. The Royal Society of Chemistry, 2016.

[28] L. A. Woolf, R. Mills, D. G. Leaist, C. Erkey, A. Akgerman, A. J. Easteal, D. G. Miller, J. G. Albright, S. F. Y. Li, and W. Wakeham, Measurement of the Transport Properties of Fluids, ch. 9 - Diffusion Coefficients, pp. 227–320. Oxford: Blackwell Science Publications, 1991.

[29] E. L. Cussler, Multicomponent diffusion. Chemical Engineering Monographs, Elsevier Scientific Publishing Company, 1976.

[30] M. Pelletier, "Quantitative analysis using Raman spectrometry," Appl. Spectrosc., vol. 57, pp. 20A–42A, 2003.

[31] J.-B. Salmon, A. Ajdari, P. Tabeling, L. Servant, D. Talaga, and M. Joanicot, "In situ Raman imaging of interdiffusion in a microchannel," Appl. Phys. Lett., vol. 86, p. 094106, 2005.

[32] A. Bardow, V. Göke, H.-J. Koß, and W. Marquardt, "Ternary diffusivities by model-based analysis of Raman spectroscopy measurements," AIChE J., vol. 52, pp. 4004–4015, 2006.

[33] G. M. Whitesides, "The origins and the future of microfluidics," Nature, vol. 442, pp. 368–373, 2006.

[34] A. Estévez-Torres, T. L. Saux, C. Gosse, A. Lemarchand, A. Bourdoncle, and L. Jullien, "Fourier transform to analyse reaction-diffusion dynamics in a microsystem," Lab Chip, vol. 8, pp. 1205–1209, 2008.

[35] K. Pappaert, J. Biesemans, D. Clicq, S. Vankrunkelsven, and G. Desmet, "Measurements of diffusion coefficients in 1-D micro- and nanochannels using shear-driven flows," Lab Chip, vol. 5, pp. 1104–1110, 2005.

[36] E. Häusler, P. Domagalski, M. Ottens, and A. Bardow, "Microfluidic diffusion measurements: The optimal H-cell," Chem. Eng. Sci., vol. 72, pp. 45–50, 2012.

[37] C. D. Costin, R. K. Olund, B. A. Staggemeier, A. K. Torgerson, and R. E. Synovec, "Diffusion coefficient measurement in a microfluidic analyzer using dual-beam microscale-refractive index gradient detection: Application to on-chip molecular size determination," J. Chromatogr. A, vol. 1013, pp. 77–91, 2013.

[38] M. H. V. Werts, V. Raimbault, R. Texier-Picard, R. Poizat, O. Francais, L. Griscom, and J. R. G. Navarro, "Quantitative full-colour transmitted light microscopy and dyes for concentration mapping and measurement of diffusion coefficients in microfluidic architectures," Lab Chip, vol. 12, pp. 808–820, 2012.

[39] E. Ryckeboer, J. Vierendeels, A. Lee, S. Werquin, P. Bienstman, and R. Baets, "Measurement of small molecule diffusion with an optofluidic silicon chip," Lab Chip, vol. 13, pp. 4392–4399, 2013.

[40] S. G. R. Lefortier, P. J. Hamersma, A. Bardow, and M. T. Kreutzer, "Rapid microfluidic screening of CO_2 solubility and diffusion in pure and mixed solvents," Lab Chip, vol. 12, pp. 3387–3391, 2012.

[41] M. S. Munson, K. R. Hawkins, M. S. Hasenbank, and P. Yager, "Diffusion based analysis in a sheath flow microchannel: the sheath flow T-sensor," Lab Chip, vol. 5, pp. 856–862, 2005.

[42] C. Peters, J. Thien, L. Wolff, H.-J. Koß, and A. Bardow, "Quaternary diffusion coefficients in liquids from microfluidics and Raman microspectroscopy: Cyclohexane + toluene+ acetone+ methanol," J. Chem. Eng. Data, 2019.

[43] J. Turner, "Multicomponent convection," Annu. Rev. of Fluid Mech., vol. 17, pp. 11–44, 1985.

[44] A. Fick, "Ueber diffusion," Ann. Phys. -Leipzig, vol. 94, pp. 59–86, 1855.

[45] X. Liu, A. Martín-Calvo, E. McGarrity, S. K. Schnell, S. Calero, J.-M. Simon, D. Bedeaux, S. Kjelstrup, A. Bardow, and T. J. H. Vlugt, "Fick diffusion coefficients in ternary liquid systems from equilibrium molecular dynamics simulations," Ind. Eng. Chem. Res., vol. 51, pp. 10247–10258, 2012.

[46] A. Bardow, E. Kriesten, M. A. Voda, F. Casanova, B. Blümich, and W. Marquardt, "Prediction of multicomponent mutual diffusion in liquids: Model discrimination using NMR data," Fluid Phase Equilibr., vol. 278, pp. 27–35, Apr. 2009.

[47] G. Guevara-Carrion, Y. Gaponenko, T. Janzen, J. Vrabec, and V. Shevtsova, "Diffusion in multi-component liquids: From microscopic to macroscopic scales," J. Phys. Chem. B, 2016.

[48] A. Heller, M. S. Fleys, J. Chen, G. P. van der Laan, M. H. Rausch, and A. P. Fröba, "Thermal and mutual diffusivity of binary mixtures of n-dodecane and n-tetracontane with carbon monoxide, hydrogen, and water from dynamic light scattering (dls)," J. Chem. Eng. Data, vol. 61, pp. 1333–1340, 2016.

[49] S. P. Cadogan, G. C. Maitland, and J. P. M. Trusler, "Diffusion coefficients of CO_2 and N_2 in water at temperatures between 298.15 K and 423.15 K at pressures up to 45 MPa," J. Chem. Eng. Data, vol. 59, pp. 519–525, 2014.

[50] B. E. Poling, J. M. Prausnitz, and J. P. O'Connell, eds., The Properties of Gases and Liquids. Digital Engineering Library @ McGraw-Hill, 5 ed., 2004.

[51] R. B. Bird, W. E. Stewart, and E. N. Lightfoot, Transport Phenomena. John Wiley & Sons, 2 ed., 2002.

[52] G. Moggridge, "Prediction of the mutual diffusivity in binary non-ideal liquid mixtures from the tracer diffusion coefficients," Chem. Eng. Sci., vol. 71, pp. 226–238, 2012.

[53] J. W. Mutoru and A. Firoozabadi, "Form of multicomponent Fickian diffusion coefficients matrix," J. Chem. Thermodyn., vol. 43, pp. 1192–1203, 2011.

[54] R. Krishna, "Uphill diffusion in multicomponent mixtures," Chem. Soc. Rev., vol. 44, pp. 2812–2836, 2015.

[55] R. Krishna and J. A. Wesselingh, "The Maxwell-Stefan approach to mass transfer," Chem. Eng. Sci., vol. 52, pp. 861–911, 1997.

[56] G. D. Kuiken, Thermodynamics of Irreversible Processes. New York: Wiley, 1994.

[57] A. Bardow, Model-based Experimental Analysis of Multicomponent Diffusion. PhD thesis, RWTH Aachen University, 2004.

[58] R. Taylor and H. A. Kooijman, "Composition derivatives of activity coefficient models (for the estimation of thermodynamic factors in diffusion)," Chem. Eng. Commun., vol. 102, pp. 87–106, 1991.

[59] H. A. Kooijman and R. Taylor, "Estimation of diffusion coefficients in multicomponent liquid systems," Ind. Eng. Chem. Res., vol. 30, pp. 1217–1222, 1991.

[60] O. O. Medvedev and A. A. Shapiro, "Modeling diffusion coefficients in binary mixtures," Fluid Phase Equilibr., vol. 225, pp. 13–22, 2004.

[61] J. Gmehling, U. Onken, W. Arlt, P. Grenzheuser, U. Weidlich, and J. R. B. Kolbe, Chemistry Data Series, Volume I, Part 1-8. DECHEMA, 1991-2014.

[62] S. K. Schnell, T. J. Vlugt, J.-M. Simon, D. Bedeaux, and S. Kjelstrup, "Thermodynamics of a small system in a μT reservoir," Chem. Phys. Lett., vol. 504, pp. 199–201, 2011.

[63] S. Deublein, B. Eckl, J. Stoll, S. V. Lishchuk, G. Guevara-Carrion, C. W. Glass, T. Merker, M. Bernreuther, H. Hasse, and J. Vrabec, "ms2: A molecular simulation tool for thermodynamic properties," Comput. Phys. Commun., vol. 182, pp. 2350–2367, 2011.

[64] G. Guevara-Carrion, T. Janzen, Y. M. Muñoz-Muñoz, and J. Vrabec, "Mutual diffusion of binary liquid mixtures containing methanol, ethanol, acetone, benzene, cyclohexane, toluene, and carbon tetrachloride," J. Chem. Phys., vol. 144, p. 124501, 2016.

[65] X. Liu, T. J. Vlugt, and A. Bardow, "Maxwell-Stefan diffusivities in liquid mixtures: Using molecular dynamics for testing model predictions," Fluid Phase Equilibr., vol. 301, pp. 110–117, 2011.

[66] X. Liu, A. Bardow, and T. H. Vlugt, "Multicomponent Maxwell-Stefan diffusivities at infinite dilution," Ind. Eng. Chem. Res., vol. 50, pp. 4776–4782, 2011.

[67] P. Tabeling, Introduction to Microfluidics. Oxford University Press, 2005.

[68] L. S. Darken, "Diffusion, mobility and their interrelation through free energy in binary metallic systems," Trans. AIME, vol. 175, pp. 184–201, 1948.

[69] X. Liu, T. J. Vlugt, and A. Bardow, "A predictive Darken equation for Maxwell-Stefan diffusivities in multicomponent mixtures supporting information," Ind. Eng. Chem. Res., vol. 50, pp. 10350–10358, 2011.

[70] C. Blesinger, P. Beumers, F. Buttler, C. Pauls, and A. Bardow, "Temperature-dependent diffusion coefficients from 1D Raman spectroscopy," J. Solution Chem., vol. 43, pp. 144–157, 2014.

[71] G. M. Kontogeorgis and R. Gani, eds., Computer Aided Property Estimation for Process and Product Design, ch. 9, pp. 205–228. Elsevier Ltd, 2004.

[72] A. Bouchaudy, C. Loussert, and J.-B. Salmon, "Steady microfluidic measurements of mutual diffusion coefficients of liquid binary mixtures," AIChE J., vol. 64, pp. 358–366, 2017.

[73] A. Bardow, V. Göke, H.-J. Koß, K. Lucas, and W. Marquardt, "Concentration-dependent diffusion coefficients from a single experiment using model-based Raman spectroscopy," Fluid Phase Equilibr., vol. 228, pp. 357–366, 2005.

[74] F. Rossi, V. K. Vanag, and I. R. Epstein, "Pentanary cross-diffusion in water-in-oil microemulsions loaded with two components of the Belousov–Zhabotinsky reaction," Chem. - Eur. J., vol. 17, pp. 2138–2145, 2011.

[75] S. A. Sanni and P. Hutchison, "Diffusivities and densities for binary liquid mixtures," J. Chem. Eng. Data, vol. 18, pp. 317–322, 1973.

[76] Y. Lin, X. Yu, Z. Wang, S.-T. Tu, and Z. Wang, "Measurement of temperature-dependent diffusion coefficients using a confocal Raman microscope with microfluidic chips considering laser-induced heating effect," Anal. Chim. Acta, vol. 667, pp. 103–112, 2010.

[77] G. Wittko and W. Köhler, "Influence of isotopic substitution on the diffusion and thermal diffusion coefficient of binary liquids," Eur. Phys. J., vol. 21, pp. 283–291, 2006.

[78] A. Mialdun and V. Shevtsova, "Temperature dependence of Soret and diffusion coefficients for toluene–cyclohexane mixture measured in convection-free environment," J. Chem. Phys., vol. 143, p. 224902, 2015.

[79] J. Winkelmann, Diffusion in Gases, Liquids and Electrolytes - Gases in Gases, Liquids and their Mixtures. Springer Materials, 2007.

[80] M. H. Rausch, L. Hopf, A. Heller, A. Leipertz, and A. P. Fröba, "Binary diffusion coefficients for mixtures of ionic liquids [EMIM][N(CN)$_2$], [EMIM][NTf$_2$], and [HMIM][NTf$_2$] with acetone and ethanol by dynamic light scattering (DLS),"

J. Phys. Chem. B, vol. 117, pp. 2429–2437, 2013.

[81] W. Krahn, G. Schwelger, and K. Lucas, "Light scattering measurements of mutual diffusion coefficients in binary liquid mixtures," J. Phys. Chem., vol. 87, pp. 4515–4519, 1983.

[82] A. Bardow, W. Marquardt, V. Göke, H.-J. Koß, and K. Lucas, "Model-based measurement of diffusion using Raman spectroscopy," AIChE J., vol. 49, pp. 323–334, 2003.

[83] H. Guo, Y. Chen, W. Lu, L. Li, and M. Wang, "In situ Raman spectroscopic study of diffusion coefficients of methane in liquid water under high pressure and wide temperatures," Fluid Phase Equilibr., vol. 360, pp. 274–278, 2013.

[84] A. Bardow, "On the interpretation of ternary diffusion measurements in low-molecular weight fluids by dynamic light scattering," Fluid Phase Equilibr., vol. 251, pp. 121–127, 2007.

[85] A. Sethy and H. T. Cullinan, "Transport of mass in ternary liquid-liquid systems. part I. diffusion studies," AIChE J., vol. 21, pp. 571–575, 1975.

[86] I. M. J. J. van de Ven-Lucassen, M. F. Kemmere, and P. J. A. M. Kerhof, "Complications in the use of the Taylor dispersion method for ternary diffusion measurements: Methanol + acetone + water mixtures," J. Solution Chem., vol. 26, pp. 1145–1167, 1997.

[87] C. Secuianu, G. C. Maitland, J. P. M. Trusler, and W. A. Wakeham, "Mutual diffusion coefficients of aqueous KCl at high pressures measured by the Taylor dispersion method," J. Chem. Eng. Data, vol. 56, pp. 4840–4848, 2011.

[88] D. G. Leaist, "Determination of ternary diffusion coefficients by the Taylor dispersion method," J. Phys. Chem., vol. 94, pp. 5180–5183, 1990.

[89] W. E. Price, "Theory of the Taylor dispersion technique for three-component-system diffusion measurements," J. Chem. Soc. Faraday Trans.I, vol. 84, pp. 2431–2439, 1988.

[90] A. Ribeiro, C. Santos, V. Lobo, and M. Esteso, "Quaternary diffusion coefficients of beta-cyclodextrin + KCl + caffeine + water at 298.15 K using a Taylor dispersion method," J. Chem. Eng. Data, vol. 55, pp. 2610–2612, 2010.

[91] C. I. Santos, M. A. Esteso, V. M. Lobo, and A. C. Ribeiro, "Multicomponent diffusion in (cyclodextrin-drug+ salt+ water) systems:{2-Hydroxypropyl-β-cyclodextrin (HP-βCD)+ KCl+ theophylline+ water}, and {β-cyclodextrin (βCD)+ KCl+ theophylline+ water}," J. Chem. Thermodyn., vol. 59, pp. 139–

143, 2013.

[92] D. G. Miller, R. Sartorio, L. Paduano, J. A. Rard, and J. G. Albright, "Effects of different sized concentration differences across free diffusion boundaries and comparison of Gouy and Rayleigh diffusion measurements using NaCl-KCl-H_2O," J. Solution Chem., vol. 25, pp. 1185–1211, 1996.

[93] D. G. Miller, "A method for obtaining multicomponent diffusion coefficients directly from Rayleigh and Gouy fringe position data," J. Phys. Chem., vol. 92, pp. 4222–4226, 1988.

[94] A. P. Fröba, Stefan Will, Y. Nagasaka, J. Winkelmann, S. Wiegand, and W. Köhler, Experimental Thermodynamics Volume IX: Advances in Transport Properties of Fluids; Chapter 2: Optical Methods, ch. 2, pp. 19–74. RSC, 2014.

[95] D. A. Ivanov and J. Winkelmann, "Measurement of diffusion in ternary liquid mixtures by light scattering technique and comparison with Taylor dispersion data," Int. J. Thermophys., vol. 29, pp. 1921–1928, 2008.

[96] O. Annunziata, D. Buzatu, and J. G. Albright, "Protein diffusion coefficients determined by macroscopic-gradient Rayleigh interferometry and dynamic light scattering.," Langmuir, vol. 21, pp. 12085–9, 2005.

[97] A. Heller, C. Giraudet, Z. A. Makrodimitri, M. S. Fleys, J. Chen, G. P. van der Laan, I. G. Economou, M. H. Rausch, and A. P. Fröba, "Diffusivities of ternary mixtures of n-alkanes with dissolved gases by dynamic light scattering," J. Phys. Chem. B, 2016.

[98] E. Kriesten, D. Mayer, F. Alsmeyer, C. Minnich, L. Greiner, and W. Marquardt, "Identification of unknown pure component spectra by indirect hard modeling," Chemometr. Intell. Lab., vol. 93, pp. 108–119, 2008.

[99] F. Alsmeyer, H.-J. Koß, and W. Marquardt, "Indirect spectral hard modeling for the analysis of reactive and interacting mixtures," Appl. Spectrosc., vol. 58, pp. 975–985, 2004.

[100] R. Swain and M. Stevens, "Raman microspectroscopy for non-invasive biochemical analysis of single cells," Biochem. Soc. T., vol. 35, pp. 544–549, 2007.

[101] P. J. Caspers, H. A. Bruining, G. J. Puppels, G. W. Lucassen, and E. A. Carter, "In vivo confocal Raman microspectroscopy of the skin: noninvasive determination of molecular concentration profiles," J. Invest. Dermatol., vol. 116, pp. 434–442, 2001.

[102] D. V. Murphy, A. C. Eckbreth, M. B. Long, and R. K. Chang, "Spatially

resolved coherent anti-Stokes Raman spectroscopy from a line across a CH_4 jet," Opt. Lett., vol. 4, pp. 167–169, 1979.

[103] C. V. Raman, "A new radiation," Indian J. Phys., vol. 2, pp. 387–398, 1928.

[104] E. Smith and G. Dent, Modern Raman spectroscopy: a practical approach. John Wiley & Sons, 2005.

[105] A. Ewinger, G. Rinke, A. Urban, and S. Kerschbaum, "In situ measurement of the temperature of water in microchannels using laser Raman spectroscopy," Chem. Eng. J., vol. 223, pp. 129–134, 2013.

[106] J. H. Hibben, "The Raman spectra of water, aqueous solutions and ice," J. Chem. Phys., vol. 5, pp. 166–172, 1937.

[107] K. Furić, I. Ciglenečki, and B. Ćosović, "Raman spectroscopic study of sodium chloride water solutions," J. Mol. Struct., vol. 550, pp. 225–234, 2000.

[108] N. J. Everall, "Confocal Raman microscopy: Performance, pitfalls, and best practice," in Pittcon - Invited Lecture at the Symposium "50 Years of SAS: Looking to the Future with Vibrational Spectroscopy", (New Orleans, Louisiana), 2008.

[109] K. Roetmann, W. Schmunk, C. S. Garbe, and V. Beushausen, "Micro-flow analysis by molecular tagging velocimetry and planar Raman-scattering," Exp. Fluids, vol. 44, pp. 419–430, 2008.

[110] G. Rinke, A. Wenka, K. Roetmann, and H. Wackerbarth, "In situ Raman imaging combined with computational fluid dynamics for measuring concentration profiles during mixing processes," Chem. Eng. J., vol. 179, pp. 338–348, 2012.

[111] J. De Gelder, K. De Gussem, P. Vandenabeele, and L. Moens, "Reference database of Raman spectra of biological molecules," J. Raman Spectrosc., vol. 38, pp. 1133–1147, 2007.

[112] E. Pretsch, T. Clerc, J. Seibl, and W. Simon, Tables of spectral data for structure determination of organic compounds. Springer Science & Business Media, 2013.

[113] J. Crank, The mathematics of diffusion. Clarendon press Oxford, 1975.

[114] L. Zhu, T. Cai, J. Huang, T. C. Stringfellow, M. Wall, and L. Yu, "Water self-diffusion in glassy and liquid maltose measured by Raman microscopy and NMR," J. Phys. Chem. B, vol. 115, pp. 5849–5855, 2011.

[115] D. Siebel, W. Schabel, and P. Scharfer, "Diffusion in quaternary polymer solutions – model development and validation," Prog. Org. Coat., vol. 110, pp. 187–

194, 2017.

[116] W. Eguchi, M. Harada, M. Adachi, M. Tanigaki, and K. Kondo, "Position-scanning spectrophotometer as a means of observing multicomponent diffusion phenomena in liquid phase," J. Chem. Eng. Jpn., vol. 17, pp. 472–477, 1984.

[117] M. Tanigaki, K. Kondo, M. Harada, and W. Eguchi, "Measurement of ternary diffusion coefficients using a position-scanning spectrophotometer," J. Phys. Chem., vol. 87, pp. 586–591, 1983.

[118] A. B. Vir, A. S. Fabiyan, J. R. Picardo, and S. Pushpavanam, "Performance comparison of liquid–liquid extraction in parallel microflows," Ind. Eng. Chem. Res., vol. 53, pp. 8171–8181, 2014.

[119] T. P. Forbes, M. S. Munson, and S. P. Forry, "Theoretical analysis of a magnetophoresis-diffusion T-sensor immunoassay," Lab Chip, vol. 13, pp. 3935–3944, 2013.

[120] R. F. Ismagilov, A. D. Stroock, P. J. A. Kenis, G. Whitesides, and H. A. Stone, "Experimental and theoretical scaling laws for transverse diffusive broadening in two-phase laminar flows in microchannels," Appl. Phys. Lett., vol. 76, pp. 2376–2378, 2000.

[121] D. Li, Encyclopedia of Microfluidics and Nanofluidics. Springer, 2008.

[122] M. Foerster, K. Lam, E. Sorensen, and A. Gavriilidis, "In situ monitoring of microfluidic distillation," Chem. Eng. J., vol. 227, pp. 13–21, 2013.

[123] C. Xu and T. Xie, "Review of microfluidic liquid–liquid extractors," Industrial & Engineering Chemistry Research, vol. 56, pp. 7593–7622, 2017.

[124] P. Gravesen, J. Branebjerg, and O. S. Jensen, "Microfluidics – a review," J. Micromech. Microeng., vol. 3, pp. 168–162, 1993.

[125] J. P. Brody and P. Yager, "Diffusion-based extraction in a microfabricated device," Sensor Actuat. A, vol. 58, pp. 13–18, 1997.

[126] S.-Y. Teh, R. Lin, L.-H. Hung, and A. P. Lee, "Droplet microfluidics," Lab Chip, vol. 8, pp. 198–220, 2008.

[127] Z. Wu, N.-T. Nguyen, and X. Huang, "Nonlinear diffusive mixing in microchannels: theory and experiments," J. Micromech. Microeng., vol. 14, pp. 604–611, 2004.

[128] D. Thuau, C. Laval, I. Dufour, P. Poulin, C. Ayela, and J.-B. Salmon, "Engineering polymer mems using combined microfluidic pervaporation and micromolding," Microsyst Nanoeng, vol. 4, p. 15, 2018.

[129] Web of Science, "Basic search on web of science core collection using keywords "microfluidic" and "diffu*"." Internet, June 29, 2017.

[130] S. Sarkar, K. Singh, V. Shankar, and K. Shenoy, "Numerical simulation of mixing at 1-1 and 1-2 microfluidic junctions," Chem. Eng. Process., vol. 85, pp. 227–240, 2014.

[131] D. Gobby, P. Angeli, and A. Gavriilidis, "Mixing characteristics of T-type microfluidic mixers," J. Micromech. Microeng., vol. 11, p. 126, 2001.

[132] T. Hotta, S. Nii, T. Yajima, and F. Kawaizumi, "Mass tansfer characteristics of a microchannel device of split-flow type," Chem. Eng. Technol., vol. 30, pp. 208–213, 2007.

[133] M. Abonnenc, J. Josserand, and H. H. Girault, "Sandwich mixer-reactor: influence of the diffusion coefficient and flow rate ratios.," Lab Chip, vol. 9, pp. 440–8, 2009.

[134] A. E. Kamholz and P. Yager, "Theoretical analysis of molecular diffusion in pressure-driven laminar flow in microfluidic channels," Biophys. J., vol. 80, pp. 155–160, 2001.

[135] Y. Zhu and Q. Fang, "Analytical detection techniques for droplet microfluidics – a review," Anal. Chim. Acta, vol. 787, pp. 24–35, 2013.

[136] B. H. Weigl and P. Yager, "Microfluidic diffusion-based separation and detection," Science, vol. 283, pp. 346–347, 1999.

[137] A. Hatch, A. E. Kamholz, K. R. Hawkins, M. S. Munson, E. A. Schilling, B. H.Weigl, and P. Yager, "A rapid diffusion immunoassay in a T-sensor," Nat. Biotechnol., vol. 19, pp. 461–465, 2001.

[138] A. E. Kamholz, B. H. Weigl, B. A. Finlayson, and P. Yager, "Quantitative analysis of molecular interaction in a microfluidic channel: The T-sensor," Anal. Chem., vol. 71, pp. 5340–5347, 1999.

[139] A. E. Kamholz and P. Yager, "Molecular diffusive scaling laws in pressure-driven microfluidic channels: deviation from one-dimensional approximations," Sensor. Actuat. B - Chem., vol. 82, pp. 117–121, 2002.

[140] C. Laval, A. Bouchaudy, and J.-B. Salmon, "Fabrication of microscale materials with programmable composition gradients," Lab Chip, vol. 16, pp. 1234–1242, 2016.

[141] C. Blesinger, S. E. Yalcin, C. Pauls, and A. Bardow, "Rapid multicomponent diffusion measurements using microfluidics," in Microfluidics 2012, 2012.

[142] T. M. Squires and S. R. Quake, "Microfluidics: Fluid physics at the nanoliter scale," Rev. Mod. Phys., vol. 77, pp. 977–1026, 2005.

[143] Z. Wu and N.-T. Nguyen, "Convective-diffusive transport in parallel lamination micromixers," Microfluid. Nanofluid., vol. 1, pp. 208–217, 2005.

[144] G. R. Wang, F. Yang, and W. Zhao, "There can be turbulence in microfluidics at low reynolds number," Lab Chip, vol. 14, pp. 1452–1458, 2014.

[145] W. Schröder, Fluidmechanik. Verlag Mainz, 2004.

[146] Q.-L. Chen, K.-J. Wu, and C.-H. He, "Investigation on liquid flow characteristics in microtubes," AIChE J., 2014.

[147] A. P. Sudarsan and V. M. Ugaz, "Fluid mixing in planar spiral microchannels," Lab Chip, vol. 6, pp. 74–82, 2006.

[148] H. Herwig, Strömungsmechanik - Einführung in die Physik von technischen Strömungen. Vieweg +Teubner Verlag, 2008.

[149] J. Dambrine, B. Géraud, and J.-B. Salmon, "Interdiffusion of liquids of different viscosities in a microchannel," New J. Phys., vol. 11, p. 075015, 2009.

[150] H. Song, Y. Wang, and K. Pant, "Cross-stream diffusion under pressure-driven flow in microchannels with arbitrary aspect ratios: a phase diagram study using a three-dimensional analytical model," Microfluid. Nanofluid., vol. 12, pp. 265–277, 2012.

[151] M. N. Kashid, A. Renken, and L. Kiwi-Minsker, "Gas–liquid and liquid–liquid mass transfer in microstructured reactors," Chem. Eng. Sci., vol. 66, pp. 3876–3897, 2011.

[152] A. J. deMello, "Control and detection of chemical reactions in microfluidic systems," Nature, vol. 442, pp. 394–402, 2006.

[153] H. Neff, A. M. N. Lima, F. C. C. L. Loureiro, and L. A. L. de Almeida, "Transient response analysis and modeling of near wall flow conditions in a micro channel: evidence of slip flow," Microfluid. Nanofluid., vol. 3, pp. 591–602, 2007.

[154] H. Murrenhoff, Grundlagen der Fluidtechnik - Hydraulik. Shaker Verlag, 2012.

[155] C. Kositanont, S. Putivisutisak, T. Tagawa, H. Yamada, and S. Assabumrungrat, "Multiphase parallel flow stabilization in curved microchannel," Chem. Eng. J., vol. 253, pp. 332–340, 2014.

[156] D. Di Carlo, "Inertial microfluidics," Lab Chip, vol. 9, pp. 3038–3046, 2009.

[157] C. Galletti, G. Arcolini, E. Brunazzi, and R. Mauri, "Mixing of binary fluids with composition-dependent viscosity in a T-shaped micro-device," Chem. Eng. Sci., vol. 123, pp. 300–310, 2015.

[158] P. J. Stiles and D. F. Fletcher, "Hydrodynamic control of the interface between two liquids flowing through a horizontal or vertical microchannel," Lab Chip, vol. 4, pp. 121–124, 2004.

[159] A. Sadeghi, "Depletion of cross-stream diffusion in the presence of viscoelasticity," AIChE J., vol. 61, pp. 4533–4541, 2015.

[160] S. K. Yoon, M. Mitchell, E. R. Choban, and P. J. A. Kenis, "Gravity-induced reorientation of the interface between two liquids of different densities flowing laminarly through a microchannel," Lab Chip, vol. 5, pp. 1259–1263, 2005.

[161] Y. Lin, X. Yu, Z. Wang, S.-T. Tu, and Z. Wang, "Laminar flow diffusion interface control in a microchannel with accurate Raman measurement," Chem. Eng. Process., vol. 57-58, pp. 1–7, 2012.

[162] J.-B. Salmon and A. Ajdari, "Transverse transport of solutes between co-flowing pressure driven streams for microfluidic studies of diffusion/reaction processes," J. Appl. Phys., vol. 101, p. 074902, 2007.

[163] Y. Lam, X. Chen, and C. Yang, "Depthwise averaging approach to cross-stream mixing in a pressure-driven microchannel flow," Microfluid. Nanofluid., vol. 1, pp. 218–226, 2005.

[164] D. A. Beard, "Taylor dispersion of a solute in a microfluidic channel," J. Appl. Phys., vol. 89, pp. 4667–4669, 2001.

[165] D. Friedrich, C. P. Please, and T. Melvin, "Design of novel microfluidic concentration gradient generators suitable for linear and exponential concentration ranges," Chem. Eng. J., vol. 193, pp. 296–303, 2012.

[166] X. He, Q. Shao, W. Kong, L. Yu, X. Zhang, and Y. Deng, "A simple method for estimating mutual diffusion coefficients of ionic liquids-water based on an optofluidic chip," Fluid Phase Equilibr., vol. 366, pp. 9–15, 2014.

[167] M. A. Holden, S. Kumar, E. T. Castellana, A. Beskok, and P. S. Cremer, "Generating fixed concentration arrays in a microfluidic device," Sensor. Actuat. B - Chem., vol. 92, pp. 199–207, 2003.

[168] D. Schafer, M. Müller, M. Bonn, D. W. M. Marr, J. van Maarseveen, and J. Squier, "Coherent anti-Stokes Raman scattering microscopy for quantitative characterization of mixing and flow in microfluidics," Opt. Lett., vol. 34,

pp. 211–213, 2009.

[169] F. Loureiro, A. B. Neto, C. Moreira, A. Lima, and H. Neff, "A method for determining the mutual diffusion coefficient of molecular solutes based on surface plasmon resonance sensing," Sensor. Actuat. B - Chem., vol. 154, pp. 129–136, 2011.

[170] A. Estévez-Torres, C. Gosse, T. L. Saux, J.-F. Allemand, V. Croquette, H. Berthoumieux, A. Lemarchand, and L. Jullien, "Fourier analysis to measure diffusion coefficients and resolve mixtures on a continuous electrophoresis chip," Anal. Chem., vol. 79, pp. 8222–8231, 2007.

[171] A. Heeren, C. Luo, G. Roth, A. Ganser, R. Brock, K.-H. Wiesmueller, W. Henschel, and D. Kern, "Diffusion along microfluidic channels," Microelectron. Eng., vol. 83, pp. 1669–1672, 2006.

[172] S. A. Sanni, C. J. D. Fell, and H. P. Hutchison, "Diffusion coefficients and densities for binary organic liquid mixtures," J. Chem. Eng. Data, vol. 16, pp. 424–427, 1971.

[173] P. Arosio, K. Hu, F. A. Aprile, T. Müller, and T. P. Knowles, "A microfluidic diffusion viscometer for rapid analysis of complex solutions," Anal. Chem., vol. 88, pp. 3488–3493, 2016.

[174] K. Carlsson and B. Karlberg, "Determination of octanol-water partition coefficients using a micro-volume liquid-liquid flow extraction system," Anal. Chim. Acta, vol. 423, pp. 137 – 144, 2000.

[175] J. Swarts, A. Janssen, and R. Boom, "Temperature effects during practical operation of microfluidic chips," Chem. Eng. Sci., vol. 63, pp. 5252–5257, 2008.

[176] D. G. Leaist, "Diffusion coefficients of four-component systems from Taylor dispersion profiles. sucrose + LiCI + KCI + water and NBu_4Br + LiBr + KBr + water," Ber. Bunsenges. Phys. Chem., vol. 95, pp. 117–122, 1991.

[177] T. Misek, R. Berger, and J. Schröter, Standard Test Systems for Liquid Extraction. European Federation of Chemical Engineering, 2nd ed., 1985.

[178] S. Enders, H. Kahl, and J. Winkelmann, "Surface tension of the ternary system water + acetone + toluene," J. Chem. Eng . Data, vol. 52, pp. 1072–1079, 2007.

[179] "Acetone dynamic viscosity: Datasheet from "dortmund data bank (ddb) – thermophysical properties edition 2014" in springermaterials (http://materials.springer.com/thermophysical/docs/vis_c4)." Copyright 2010-2014 Springer-Verlag Berlin Heidelberg & DDBST GmbH, Oldenburg,

Germany.

[180] "Acetone thermal conductivity: Datasheet from "Dortmund Data Bank (DDB) – thermophysical properties edition 2014" in springermaterials (http://materials.springer.com/thermophysical/docs/tcn_c4)." Copyright 2010-2014 Springer-Verlag Berlin Heidelberg & DDBST GmbH, Oldenburg, Germany.

[181] M. F. Bolotnikov, Y. A. Neruchev, and O. S. Ryshkova, "Density of some 1-chloroalkanes within the temperature range from (253.15 to 423.15) K," J. Chem. Eng. Data, vol. 52, pp. 2514–2516, 2007.

[182] C. Wohlfarth, "Viscosity of 1-chlorobutane: Datasheet from Landolt-Börnstein - group IV physical chemistry · volume 25: "supplement to iv/18" in springermaterials (http://dx.doi.org/10.1007/978-3-540-75486-2_109)." Copyright 2008 Springer-Verlag Berlin Heidelberg.

[183] M. Kleiber, R. Joh, and R. Span, "D3 properties of pure fluid substances: Datasheet from VDI-Buch · volume : "VDI heat atlas" in springermaterials (http://dx.doi.org/10.1007/978-3-540-77877-6_18)." Copyright 2010 Springer-Verlag.

[184] S. A. Beg, N. M. Tukur, D. K. Al-Harbi, and E. Z. Hamad, "Saturated liquid densities of benzene, cyclohexane, and hexane from 298.15 to 473.15 K," J. Chem. Eng. Data, vol. 38, pp. 461–464, 1993.

[185] "Cyclohexane dynamic viscosity: Datasheet from "Dortmund Data Bank (DDB) – thermophysical properties edition 2014" in springermaterials (http://materials.springer.com/thermophysical/docs/vis_c50)." Copyright 2010-2014 Springer-Verlag Berlin Heidelberg & DDBST GmbH, Oldenburg, Germany.

[186] "Cyclohexane thermal conductivity: Datasheet from "dortmund data bank (ddb) – thermophysical properties edition 2014" in springermaterials (http://materials.springer.com/thermophysical/docs/tcn_c50)." Copyright 2010-2014 Springer-Verlag Berlin Heidelberg & DDBST GmbH, Oldenburg, Germany.

[187] D. C. Landaverde-Cortes, A. Estrada-Baltazar, G. A. Iglesias-Silva, and K. R. Hall, "Densities and viscosities of MTBE+ heptane or octane at p= 0.1 MPa from (273.15 to 363.15) K," J. Chem. Eng. Data, vol. 52, pp. 1226–1232, 2007.

[188] C. Wohlfarth, "Viscosity of heptane: Datasheet from Landolt-Börnstein - group IV physical chemistry · volume 25: "supplement to iv/18" in springerma-

terials (http://dx.doi.org/10.1007/978-3-540-75486-2_269)." Copyright 2008 Springer-Verlag Berlin Heidelberg.

[189] "Heptane thermal conductivity: Datasheet from "Dortmund Data Bank (DDB) – thermophysical properties edition 2014" in springermaterials (http://materials.springer.com/thermophysical/docs/tcn_c91)." Copyright 2010-2014 Springer-Verlag Berlin Heidelberg & DDBST GmbH, Oldenburg, Germany.

[190] W. Fan, Q. Zhou, S. Zhang, and R. Yan, "Excess molar volume and viscosity deviation for the methanol+ methyl methacrylate binary system at T=(283.15 to 333.15) K," J. Chem. Eng. Data, vol. 53, pp. 1836–1840, 2008.

[191] "Methanol dynamic viscosity: Datasheet from "dortmund data bank (ddb) – thermophysical properties edition 2014" in springermaterials (http://materials.springer.com/thermophysical/docs/vis_c110)." Copyright 2010-2014 Springer-Verlag Berlin Heidelberg & DDBST GmbH, Oldenburg, Germany.

[192] "Methanol thermal conductivity: Datasheet from "Dortmund Data Bank (DDB) – thermophysical properties edition 2014" in springermaterials (http://materials.springer.com/thermophysical/docs/tcn_c110)." Copyright 2010-2014 Springer-Verlag Berlin Heidelberg & DDBST GmbH, Oldenburg, Germany.

[193] E. Vercher, A. V. Orchilles, P. J. Miguel, and A. Martínez-Andreu, "Volumetric and ultrasonic studies of 1-ethyl-3-methylimidazolium trifluoromethanesulfonate ionic liquid with methanol, ethanol, 1-propanol, and water at several temperatures," J. Chem. Eng. Data, vol. 52, pp. 1468–1482, 2007.

[194] "1-propanol dynamic viscosity: Datasheet from "Dortmund Data Bank (DDB) – thermophysical properties edition 2014" in springermaterials (http://materials.springer.com/thermophysical/docs/vis_c140)." Copyright 2010-2014 Springer-Verlag Berlin Heidelberg & DDBST GmbH, Oldenburg, Germany.

[195] "1-propanol thermal conductivity: Datasheet from "Dortmund Data Bank (DDB) – thermophysical properties edition 2014" in springermaterials (http://materials.springer.com/thermophysical/docs/tcn_c140)." Copyright 2010-2014 Springer-Verlag Berlin Heidelberg & DDBST GmbH, Oldenburg, Germany.

[196] M. Chorazewski, J.-P. E. Grolier, and S. L. Randzio, "Isobaric thermal expan-

sivities of toluene measured by scanning transitiometry at temperatures from (243 to 423) K and pressures up to 200 MPa," J. Chem. Eng. Data, vol. 55, pp. 5489–5496, 2010.

[197] "Toluene thermal conductivity: Datasheet from "Dortmund Data Bank (DDB) – thermophysical properties edition 2014" in springermaterials (http://materials.springer.com/thermophysical/docs/tcn_c161)." Copyright 2010-2014 Springer-Verlag Berlin Heidelberg & DDBST GmbH, Oldenburg, Germany.

[198] "Toluene dynamic viscosity: Datasheet from "dortmund data bank (ddb) – thermophysical properties edition 2014" in springermaterials (http://materials.springer.com/thermophysical/docs/vis_c161)." Copyright 2010-2014 Springer-Verlag Berlin Heidelberg & DDBST GmbH, Oldenburg, Germany.

[199] "Water dynamic viscosity: Datasheet from "dortmund data bank (ddb) – thermophysical properties edition 2014" in springermaterials (http://materials.springer.com/thermophysical/docs/vis_c174)." Copyright 2010-2014 Springer-Verlag Berlin Heidelberg & DDBST GmbH, Oldenburg, Germany.

[200] "Water thermal conductivity: Datasheet from "dortmund data bank (ddb) – thermophysical properties edition 2014" in springermaterials (http://materials.springer.com/thermophysical/docs/tcn_c174)." Copyright 2010-2014 Springer-Verlag Berlin Heidelberg & DDBST GmbH, Oldenburg, Germany.

[201] "Cyclohexane-toluene excess volume: Datasheet from "dortmund data bank (ddb) – thermophysical properties edition 2014" in springermaterials (http://materials.springer.com/thermophysical/docs/ve0_c50c161)." Copyright 2010-2014 Springer-Verlag Berlin Heidelberg & DDBST GmbH, Oldenburg, Germany.

[202] "Cyclohexane-toluene enthalpy of mixing: Datasheet from "dortmund data bank (ddb) – thermophysical properties edition 2014" in springermaterials (http://materials.springer.com/thermophysical/docs/he_c50c161)." Copyright 2010-2014 Springer-Verlag Berlin Heidelberg & DDBST GmbH, Oldenburg, Germany.

[203] "Cyclohexane-toluene azeotropic data: Datasheet from "dortmund data bank (ddb) – thermophysical properties edition 2014" in springermaterials

(http://materials.springer.com/thermophysical/docs/azd_c50c161)." Copyright 2010-2014 Springer-Verlag Berlin Heidelberg & DDBST GmbH, Oldenburg, Germany.

[204] R. W. Berg, S. Brunsgaard Hansen, A. A. Shapiro, and E. H. Stenby, "Diffusion measurements in binary liquid mixtures by Raman spectroscopy," Appl. Spectrosc., vol. 61, pp. 367–373, 2007.

[205] "Cyclohexane-methanol excess volume: Datasheet from "dortmund data bank (ddb) – thermophysical properties edition 2014" in springermaterials (http://materials.springer.com/thermophysical/docs/ve0_c50c110)." Copyright 2010-2014 Springer-Verlag Berlin Heidelberg & DDBST GmbH, Oldenburg, Germany.

[206] "Cyclohexane-methanol enthalpy of mixing: Datasheet from "dortmund data bank (ddb) – thermophysical properties edition 2014" in springermaterials (http://materials.springer.com/thermophysical/docs/he_c50c110)." Copyright 2010-2014 Springer-Verlag Berlin Heidelberg & DDBST GmbH, Oldenburg, Germany.

[207] "Cyclohexane-methanol azeotropic data: Datasheet from "dortmund data bank (ddb) – thermophysical properties edition 2014" in springermaterials (http://materials.springer.com/thermophysical/docs/azd_c50c110)." Copyright 2010-2014 Springer-Verlag Berlin Heidelberg & DDBST GmbH, Oldenburg, Germany.

[208] T. Tominaga, S. Tenma, and H. Watanabe, "Diffusion of cyclohexane and cyclopentane derivatives in some polar and non-polar solvents. Effect of intermolecular and intramolecular hydrogen-bonding interactions," J. Chem. Soc., Faraday Trans., vol. 92, pp. 1863–1867, 1996.

[209] "Acetone-toluene excess volume: Datasheet from "Dortmund Data Bank (DDB) – thermophysical properties edition 2014" in springermaterials (http://materials.springer.com/thermophysical/docs/ve0_c4c161)." Copyright 2010-2014 Springer-Verlag Berlin Heidelberg & DDBST GmbH, Oldenburg, Germany.

[210] "Acetone-toluene enthalpy of mixing: Datasheet from "dortmund data bank (ddb) – thermophysical properties edition 2014" in springermaterials (http://materials.springer.com/thermophysical/docs/he_c4c161)." Copyright 2010-2014 Springer-Verlag Berlin Heidelberg & DDBST GmbH, Oldenburg, Germany.

[211] "Acetone-toluene azeotropic data: Datasheet from "Dortmund Data Bank (DDB) – thermophysical properties edition 2014" in springermaterials (http://materials.springer.com/thermophysical/docs/azd_c4c161)." Copyright 2010-2014 Springer-Verlag Berlin Heidelberg & DDBST GmbH, Oldenburg, Germany.

[212] W. Baldauf and H. Knapp, "Experimental determination of diffusion coefficients, viscosities, densities and refractive indexes of 11 binary liquid systems," Ber. Bunsen Gesell., vol. 87, pp. 304–309, 1983.

[213] "Acetone-water excess volume: Datasheet from "Dortmund Data Bank (DDB) – thermophysical properties edition 2014" in springermaterials (http://materials.springer.com/thermophysical/docs/ve0_c4c174)." Copyright 2010-2014 Springer-Verlag Berlin Heidelberg & DDBST GmbH, Oldenburg, Germany.

[214] "Acetone-water enthalpy of mixing: Datasheet from "dortmund data bank (ddb) – thermophysical properties edition 2014" in springermaterials (http://materials.springer.com/thermophysical/docs/he_c4c174)." Copyright 2010-2014 Springer-Verlag Berlin Heidelberg & DDBST GmbH, Oldenburg, Germany.

[215] "Acetone-water azeotropic data: Datasheet from "Dortmund Data Bank (DDB) – thermophysical properties edition 2014" in springermaterials (http://materials.springer.com/thermophysical/docs/azd_c4c174)." Copyright 2010-2014 Springer-Verlag Berlin Heidelberg & DDBST GmbH, Oldenburg, Germany.

[216] D. K. Anderson, J. R. Hall, and A. L. Babb, "Mutual diffusion in non-ideal binary liquid mixtures," J. Phys. Chem., vol. 62, pp. 404–408, 1958.

[217] S. Rehfeldt, Mehrkomponentendiffusion in Flüssigkeiten. PhD thesis, München, Techn. Univ., Diss., 2009.

[218] M. T. Tyn and W. F. Calus, "Temperature and concentration dependence of mutual diffusion coefficients of some binary liquid systems," J. Chem. Eng. Data, vol. 20, pp. 310–316, 1975.

[219] T. Janzen, S. Zhang, A. Mialdun, G. Guevara-Carrion, J. Vrabec, M. He, and V. Shevtsova, "Mutual diffusion governed by kinetics and thermodynamics in the partially miscible mixture methanol+ cyclohexane," Phys. Chem. Chem. Phys., vol. 19, pp. 31856–31873, 2017.

[220] Z. Atik and W. Kerboub, "Liquid-liquid equilibrium of (cyclohexane + 2,2,2-

trifluoroethanol) and (cyclohexane + methanol) from (278.15 to 318.15) K," J. Chem. Eng. Data, vol. 53, pp. 1669–1671, 2008.

[221] I. Nagata, "Liquid-liquid equilibria for four ternary systems containing methanol and cyclohexane," Fluid Phase Equilibr., vol. 18, pp. 83–92, 1984.

[222] T. Großmann and J. Winkelmann, "Ternary diffusion coefficients of cyclohexane + toluene + methanol by Taylor dispersion measurements at 298.15 K. part 1. toluene-rich area," J. Chem. Eng. Data, vol. 54, pp. 405–410, 2009.

[223] V. Göke, Messung von Diffusionskoeffizienten mittels eindimensionaler Ramanspektroskopie. PhD thesis, RWTH Aachen University, 2005.

[224] T. Janzen and J. Vrabec, "Diffusion coefficients of a highly non-ideal ternary liquid mixture: Cyclohexane+ toluene+ methanol," Industrial & Engineering Chemistry Research, 2018.

[225] J. Bulieka and J. Prochazka, "Diffusion coefficients in some ternary systems," J. Chem. Eng. Data, vol. 21, pp. 452–456, 1976.

[226] G. Guevara-Carrion, C. Nieto-Draghi, J. Vrabec, and H. Hasse, "Prediction of transport properties by molecular simulation: Methanol and ethanol and their mixture," J. Phys. Chem. B, vol. 112, pp. 16664–16674, 2008.

[227] L. Wolff, P. Zangi, T. Brands, M. H. Rausch, H.-J. Koß, A. P. Fröba, and A. Bardow, "Concentration-dependent diffusion coefficients of binary gas mixtures using a loschmidt cell with holographic interferometry," International Journal of Thermophysics, vol. 39, p. 133, Oct 2018.

[228] W. Rohsenow, J. P. Hartnett, and Y. I. Cho, Handbook of Heat Transfer. McGraw-Hill Professional, third ed., 1998.

[229] P. Beumers, T. Brands, H.-J. Koss, and A. Bardow, "Model-free calibration of Raman measurements of reactive systems: Application to monoethanolamine/water/CO_2," Fluid Phase Equilibr., vol. 424, pp. 52–57, 2016.

[230] J. P. O'Connell and J. M. Haile, Thermodynamics: Fundamentals for applications. Cambridge University Press, 2005.

[231] M. Kleiner and G. Sadowski, "Modelling of polar systems using PCP-SAFT: An approach to account for induced-association interactions," J. Phys. Chem. C, vol. 111, pp. 15544–53, 2007.

[232] J. Gross and G. Sadowski, "Perturbed-chain SAFT: An equation of state based on a perturbation theory for chain molecules," Ind. Eng. Chem. Res., vol. 40,

pp. 1244–1260, 2001.

[233] J. Gross and G. Sadowski, "Application of the perturbed-chain SAFT equation of state to associating systems," Ind. Eng. Chem. Res., vol. 41, pp. 5510–5515, 2002.

[234] J. Gross and J. Vrabec, "An equation of state contribution for polar components: Dipolar molecules," AIChE J., vol. 52, pp. 1194–1201, 2006.

[235] S. Haase, "Validierung eines mikrofluidischen Versuchsaufbaus für Diffusion in Flüssigkeiten mit konfokaler Ramanspektroskopie," Masterarbeit, RWTH Aachen University, 2015.

[236] A. Juring, "Validierung eines optimierten mikrofluidischen Messverfahrens für die Bestimmung von multikomponenten Diffusionskoeffizienten in Flüssigkeiten," Studienarbeit, RWTH Aachen University, 2013.

[237] R. K. Ghai and F. A. L. Dullien, "Diffusivities and viscosities of some binary liquid nonelectrolytes at 25.deg.," J. Phys. Chem., vol. 78, pp. 2283–2291, 1974.

[238] M. Claessens, p. Fiasse, C. Fabre, D. Zimmermann, and J. Reisse, "Experimental and theoretical study of diffusion in pure liquids and solutions," Nouv. J. Chim., vol. 8, pp. 357–363, 1984.

[239] J. Lewis, "Some determinations of liquid-phase diffusion coefficlents by means of an improved diaphragm cell," J. Appl. Chem., vol. 5, pp. 228–237, 1955.

[240] N. von Solms, I. A. Kouskoumvekaki, M. L. Michelsen, and G. M. Kontogeorgis, "Capabilities, limitations and challenges of a simplified PC-SAFT equation of state," Fluid Phase Equilibr., vol. 241, pp. 344–353, 2006.

[241] S. Rehfeldt and J. Stichlmair, "Measurement and prediction of multicomponent diffusion coefficients in four ternary liquid systems," Fluid Phase Equilibr., vol. 290, pp. 1–14, 2010.

[242] T. Großmann and J. Winkelmann, "Ternary diffusion coefficients of glycerol + acetone + water by Taylor dispersion measurements at 298.15 K," J. Chem. Eng. Data, vol. 50, pp. 1396–1403, 2005.

[243] P. Beumers, D. Engel, T. Brands, H.-J. Koß, and A. Bardow, "Robust analysis of spectra with strong background signals by first-derivative indirect hard modeling (fd-ihm)," Chemometrics and Intelligent Laboratory Systems, vol. 172, pp. 1 – 9, 2018.

[244] D. G. Miller, J. G. Albright, R. Mathew, C. M. Lee, J. A. Rard, and L. B. Eppstein, "Isothermal diffusion coefficients of sodium chloride-magnesium chloride-

water at 25 °C. 5. Solute concentration ratio of 1:1 and some Rayleigh results," J. Phys. Chem., vol. 97, pp. 3885–3899, 1993.

[245] M. Petrowsky and R. Frech, "Application of the compensated Arrhenius formalism to self-diffusion: Implications for ionic conductivity and dielectric relaxation," The Journal of Physical Chemistry B, vol. 114, pp. 8600–8605, 2010.

[246] J. W. Moore and R. M. Wellek, "Diffusion coefficients of n-heptane and n-decane in n-alkanes and n-alcohols at several temperatures," J. Chem. Eng. Data, vol. 19, pp. 136–140, 1974.

[247] L. W. Shemilt and R. Nagarajan, "Liquid diffusivities for the system methanol-toluene," Can. J. Chem., vol. 45, pp. 1143–1148, 1967.

[248] F. P. Lees and P. Sarram, "Diffusion coefficient of water in some organic liquids," J. Chem. Eng. Data, vol. 16, pp. 41–44, 1971.

[249] L. Bonoli and P. Witherspoon, "Diffusion of aromatic and cycloparaffin hydrocarbons in water from 2 to 60. deg.," J. Phys. Chem., vol. 72, pp. 2532–2534, 1968.

[250] C. Blesinger, A. Juring, C. Pauls, and A. Bardow, "Effiziente Messung von Diffusionskoeffizienten durch Mikrofluidik und modellgestützte experimentelle Analyse," in Jahrestreffen Reaktionstechnik 2014 zusammen mit der Fachgruppe Mikroreaktionstechnik, Würzburg, 28.-30.04.2014, 2014.

[251] C. Peters, J. Thien, S. Haase, C. Flake, L. Wolff, H.-J. Koß, and A. Bardow, "Microfluidic measurements of diffusion in liquids using confocal Raman spectroscopy," in 14th International Conference on Properties and Phase Equilibria for Product and Process Design (PPEPPD 2016), Porto, Portugal, 22.-26. Mai 2016, 2016.

[252] J. Thien, C. Peters, S. Haase, L. Wolff, T. Brands, H.-J. Koß, and A. Bardow, "Bestimmung von Diffusionskoeffizienten mit Mikrofluidik und konfokaler Raman-Spektroskopie," in Thermodynamik-Kolloquium 2015 vom 5.-7. Oktober 2015, Ruhr-Universität Bochum, 2015.

[253] C. A. Lieber and A. Mahadevan-Jansen, "Automated method for subtraction of fluorescence from biological Raman spectra," Appl. Spectrosc., vol. 57, pp. 1363–1367, 2003.

[254] D. Wei, S. Chen, and Q. Liu, "Review of fluorescence suppression techniques in Raman spectroscopy," Appl. Spectrosc. Rev., vol. 50, pp. 387–406, 2015.

[255] A. O'Grady, A. C. Dennis, D. Denvir, J. J. McGarvey, and S. E. Bell, "Quan-

titative Raman spectroscopy of highly fluorescent samples using pseudosecond derivatives and multivariate analysis," Anal. Chem., vol. 73, pp. 2058–2065, 2001.

[256] P. Beumers, Physically Based Models for the Analysis of Raman Spectra. PhD thesis, RWTH Aachen University, submitted 2018.

[257] J. B. Cooper, K. L. Wise, J. Groves, and W. T. Welch, "Determination of octane numbers and reid vapor pressure of commercial petroleum fuels using FT-Raman spectroscopy and partial least-squares regression analysis," Anal. Chem., vol. 67, pp. 4096–4100, 1995.

[258] A. Perro, G. Lebourdon, S. Henry, S. Lecomte, L. Servant, and S. Marre, "Combining microfluidics and FT-IR spectroscopy: towards spatially resolved information on chemical processes," React. Chem. Eng., vol. 1, pp. 577–594, 2016.

[259] G. Münch, N. Henn, and J. Laker, "Optimierung eines Versuchsaufbaus zur Messung von Flüssig-flüssig-Gleichgewichten mittels Mikrofluidik und Raman Mikro-Spektroskopie durch die Integration eines Mikromischers," Projektarbeit, RWTH Aachen University, 2017.

[260] C. Balcells, I. Pastor, E. Vilaseca, S. Madurga, M. Cascante, and F. Mas, "Macromolecular crowding effect upon in vitro enzyme kinetics: Mixed activation–diffusion control of the oxidation of NADH by pyruvate catalyzed by lactate dehydrogenase," J. Phys. Chem. B, vol. 118, pp. 4062–4068, 2014.

[261] D. Schafer, J. A. Squier, J. v. Maarseveen, D. Bonn, M. Bonn, and M. Müller, "In situ quantitative measurement of concentration profiles in a microreactor with submicron resolution using multiplex CARS microscopy," J. Am. Chem. Soc., vol. 130, pp. 11592–11593, 2008.

[262] J. Thien, "Experimentelle Bestimmung des Flüssig-flüssig-Gleichgewichts und der Diffusionskoeffizienten eines ternären Systems mit konfokaler Raman-Spektroskopie," Masterarbeit, RWTH Aachen University, 2015.

[263] J. Hansson, J. M. Karlsson, T. Haraldsson, H. Brismar, W. van der Wijngaart, and A. Russom, "Inertial microfluidics in parallel channels for high-throughput applications," Lab Chip, vol. 12, pp. 4644–4650, 2012.

[264] G. A. Cooksey, C. G. Sip, and A. Folch, "A multi-purpose microfluidic perfusion system with combinatorial choice of inputs, mixtures, gradient patterns, and flow rates," Lab Chip, vol. 9, pp. 417–426, 2009.

[265] A. Vignes, "Diffusion in binary solutions. variation of diffusion coefficient with

composition," Ind. Eng. Chem. Fund., vol. 5, pp. 189–199, 1966.

[266] J. Wesselingh and R. Krishna, Elements of mass transfer. TU Delft, Faculty of Chemical Engeneering and Materials Science Ellis Hoewood: Chichester, 1989.

[267] R. Krishna and J. M. van Baten, "The Darken relation for multicomponent diffusion in liquid mixtures of linear alkanes: An investigation using molecular dynamics (MD) simulations," Ind. Eng. Chem. Res., vol. 44, pp. 6939–6947, 2005.

[268] S. Rehfeldt and J. Stichlmair, "Measurement and calculation of multicomponent diffusion coefficients in liquids," Fluid Phase Equilibr., vol. 256, pp. 99–104, 2007.

[269] M. T. Tyn, "Estimation of diffusion coefficients of liquid mixtures at any temperature," Chem. Eng. J., vol. 12, pp. 149–150, 1976.

[270] V. Sanchez, H. Oftadeh, C. Durou, and J. P. Hot, "Restricted diffusion in binary organic liquid mixtures," J. Chem. Eng. Data, vol. 22, pp. 123–125, 1977.

[271] D. Ciceri, J. M. Perera, and G. W. Stevens, "A study of molecular diffusion across a water/oil interface in a Y–Y shaped microfluidic device," Microfluid. Nanofluid., vol. 11, pp. 593–600, 2011.

[272] A. A. Silva, R. A. Reis, and M. L. L. Paredes, "Density and viscosity of decalin, cyclohexane, and toluene binary mixtures at (283.15, 293.15, 303.15, 313.15, and 323.15) K," J. Chem. Eng. Data, vol. 54, pp. 2067–2072, 2009.

Aachener Beiträge zur Technischen Thermodynamik

ABTT 1
Philip Voll
Automated Optimization-Based Synthesis of Distributed Energy Supply Systems
1. Auflage 2014
ISBN 978-3-86130-474-6

ABTT 2
Johannes Jung
Comparative Life Cycle Assessment of Industrial Multi-Product Processes
1. Auflage 2014
ISBN 978-3-86130-471-5

ABTT 3
Franz Lanzerath
Modellgestützte Entwicklung von Adsorptionswärmepumpen
1. Auflage 2014
ISBN 978-3-86130-472-2

ABTT 4
Thorsten Brands
Einfluss der Gemischzusammensetzung auf die Verbrennung im Diesel- und GCAI-Motor
1. Auflage 2014
ISBN 978-3-95886-006-3

ABTT 5
Dominique Dechambre
Efficient Measurement of Liquid-Liquid Equilibria using Automation and Optimal Experimental Design
1. Auflage 2016
ISBN 978-395886-077-3

ABTT 6
Niklas von der Aßen
From Life-Cycle Assesement towards life-Cycle Design of Carbon Dioxide Capture and Utilization
1. Auflage 2016
ISBN 978-3-95886-080-3

ABTT 7
Matthias Lampe
Integrated Process and Organic Rankine Cycle Working Fluid Design in the Continuous-Molecular Targeting Framework
1. Auflage 2016
ISBN 978-3-95886-086-5

ABTT 8
Thomas Hülser
Optische Untersuchung der Zündvorgänge und deren Auswirkung auf die Verbrennung in PKW-Motoren
1. Auflage 2016
ISBN 978-3-95886-090-2

Aachener Beiträge zur Technischen Thermodynamik

ABTT 9
Malte Döntgen
Reaction Models from Reactive Molecular Dynamics and High-Level Kinetics Predictions
1. Auflage 2016
ISBN 978-3-95886-156-5

ABTT 10
Heike Schreiber
Experiments and Validated Models for Adsorption Thermal Energy Storage in Industrial and Residential Application
1. Auflage 2017
ISBN 978-3-95886-178-7

ABTT 11
André Dirk Sternberg
System-Wide Perspective for Life Cycle Assesment of CO_2-based C1-Chemicals
1. Auflage 2017
ISBN 978-3-95886-193-0

ABTT 12
Uwe Bau
From Dynamic Simulation to Optimal Design and Control of Adsorption Energy Systems
1. Auflage 2018
ISBN 978-3-95886-216-6

ABTT 13
Christian Jens
Modellbasiertes Design von Produkt, Lösungsmittel und Prozess für die Ameisensäure-synthese aus CO_2 und H_2
1. Auflage 2018
ISBN 978-3-95886-231-9

ABTT 14
Jan David Scheffczyk
Integrated Computer-Aided Design of Molecules and Processes using COSMO-RS
1. Auflage 2018
ISBN 978-3-95886-236-4

ABTT 15
Björn Bahl
Optimization-Based Synthesis of Large-Scale Energy Systems by Time-Series Aggregation
1. Auflage 2018
ISBN 978-3-95886-240-1

Aachener Beiträge zur Technischen Thermodynamik

ABTT 16
Bastian Liebergesell
A Milliliter-Scale Setup for the Efficient Characterization of Multicomponent Vapor-Liquid Equilibria Using Raman Spectroscopy
1. Auflage 2018
ISBN 978-3-95886-247-0

ABTT 17
Stefan Wilhelm Graf
A Design Approach for Adsorption Energy Systems Integrating Dynamic Modeling with Small-Scale Experiments
1. Auflage 2018
ISBN 978-3-95886-258-6

ABTT 18
Sebastian Kaminski
Quantum-Mechanics-Based Prediction of SAFT Parameters for Non-Associating and Associating Molecules Containing Carbon, Hydrogen, Oxygen and Nitrogen
1. Auflage 2019
ISBN 978-3-95886-270-8

ABTT 19
Maike Renate Hennen
Decision Support for the Synthesis of Energy Systems by Analysis of the Near-Optimal Solution Space
1. Auflage 2019
ISBN 978-3-95886-277-7

ABTT 20
Peyman Yamin
COSMO-RS-Based Methods for Improved Modelling of Complex Chemical Systems
1. Auflage 2019
ISBN 978-3-95886-288-3

ABTT 21
Meltem Erdogan
Assessement of Adsorbents for Drying by Experiments and Dynamic Simulations
1. Auflage 2019
ISBN 978-3-95886-303-3

ABTT 22
Christian Schulz
SRS/LIF-Messungen zur Charakterisierung rußarmer dieselähnlicher Flammen von alternativen Kraftstoffen und n-Heptan
1. Auflage 2019
ISBN 978-3-91886-310-1

Aachener Beiträge zur Technischen Thermodynamik

ABTT 23
Peter Beumers
Physically-Based Models for the Analysis of Raman Spectra
1. Auflage 2019
ISBN 978-3-95886-319-4

ABTT 24
Arne Kätelhön
Technology Choice Model for Consequential Life Cycle Assessment
1. Auflage 2019
ISBN 978-3-95886-324-8

ABTT 25
Christine Peters
Measurement of Multicomponent Diffusion in Liquids Using Raman Microspectroscopy and Microfluidics
1. Auflage 2020
ISBN 978-3-95886-337-8

Aachener Beiträge zur Technischen Thermodynamik